RTI Strategies that Work in the K-2 Classroom

Eli Johnson
and
Michelle Karns

New York London

First published 2010 by Eye On Education

Published 2013 by Routledge
711 Third Avenue, New York, NY, 10017, USA
2 Park Square, Milton Park, Abingdon, Oxon OX14 4RN

Routledge is an imprint of the Taylor & Francis Group, an informa business

Library of Congress Cataloging-in-Publication Data

Johnson, Eli R.
RTI strategies that work in the K-2 classroom / by Eli Johnson.
p. cm.
ISBN 978-1-59667-171-3
1. Remedial teaching. 2. Language arts (Early childhood)
3. Mathematics--Study and teaching (Early childhood) I. Title.
LB1029.R4J63 2010
371.102--dc22

2010036703

ISBN: 978-1-59667-171-3 (pbk)

Also Available from Eye On Education

Response to Intervention and Continuous School Improvement: Using Data, Vision, and Leadership to Design, Implement, and Evaluate a Schoolwide Prevention System
Victoria L. Bernhardt and Connie L. Hebert

Transforming High Schools through Response to Intervention (RtI): Lessons Learned and a Pathway Forward
Jeremy Koselak

Questions and Answers About RTI: A Guide to Success
Heather Moran and Anthony Petruzzelli

Math Intervention: Building Number Power with Formative Assessments, Differentiation, and Games (Grades PreK-2)
Jennifer Taylor-Cox

Math Intervention: Building Number Power with Formative Assessments, Differentiation, and Games (Grades 3-5)
Jennifer Taylor-Cox

Test Less, Assess More: A K-8 Guide to Formative Assessment
Leighangela Brady and Lisa McColl

Active Literacy Across the Curriculum
Heidi Hayes-Jacobs

Classroom Motivation from A to Z: How to Engage Your Students in Learning
Barbara R. Blackburn

Teaching, Learning, and Assessment Together: Reflective Assessments for Elementary Classrooms
Arthur K. Ellis

Differentiating by Readiness: Strategies and Lesson Plans for Tiered Instruction Grades K-8
Joni Turville, Linda Allen, and Leann Nickelsen

How the Best Teachers Avoid the 20 Most Common Teaching Mistakes
Elizabeth Breaux

How To Reach and Teach All Students—Simplified
Elizabeth Breaux

Contents

RTI Strategies that Work for K-2 Classrooms
Strategy Matrix

About the Authors

Eli Johnson is currently the Chief Academic Officer for the Golden Plains Unified School District. He provides trainings for school districts and regional organizations throughout the United States. He has previously served as an Educational Consultant for the California Department of Education, Assistant Superintendent of Curriculum and Instruction, High School Principal, Assistant Principal, and Classroom Teacher. His schools have made the highest annual point gains in the state of California and have been awarded the U.S. News & World Report Medal Honors. Eli received his undergraduate degree in education from Brigham Young University (Go Cougars!) and his graduate degree in education from the University of Washington (Go Huskies!). He is the author of *Academic Language! Academic Literacy!* Eli is married to his wonderful wife Shaunna, and they are the parents of five children: Natalie, Mikaila, Bryce, Erica, and Benjamin.

Michelle Karns is an educational consultant with over 30 years experience working with struggling learners and students impacted by adversity. She strives to develop remedies to answer their individual learning, instructional, organizational, and community needs. Michelle is committed to the realization of successful outcomes for every child. She has the unique ability to translate complex theory into easy-to use, meaningful techniques and applications for classroom settings. Working with students, administrators, and teachers in districts throughout the United States and Canada, Michelle helps create the conditions for *all* students to learn and make academic success a reality. An author of several books and multiple educational reform articles for the Association of California School Administrator's *Leadership Magazine*, she has helped thousands of students and teachers build positive relationships and meet their academic and personal success goals.

Foreword

We are eternally grateful for those who intervene in emergencies: the doctors in an ER who use their skills, knowledge, and compassion to save the infirm; the brave firemen—those immediate responders who endanger their own lives to save not only our precious possessions but also our lives; lifeguards who have the courage to fearlessly come to rescue us when we're in harm's way.

Communities depend on these brave heroes to come to our salvation when we are in need. Thankfully, they are prepared to be alert to signs of distress, to diagnose the problem, and then to employ a range of interventions including tools, techniques, strategies, and maneuvers—to provide immediate resuscitation.

And that is what this book is about, only it is located in classroom settings. Compassionate, altruistic, dedicated teachers also are alert to signs that their students may need rescuing from physiological, environmental, impoverished, or linguistic conditions that endanger their success not only in school but also throughout their future lives.

Just as these emergency workers are trained to skillfully intervene, so too are teachers trained to intervene when it becomes evident that a student is in danger of failure. The authors provide a wide range of practical tools for teachers to employ when those telltale signs of impending failure become evident.

Not all children, however, need the same interventions. Some cases are mild; others may be critical, requiring intensive care. The authors wisely provide a scale for selecting appropriate interventions. Most students (80%) can learn well using normal classroom instruction and interventions; others (15%) may need special targeted attention; and yet others (5%) will need intensive interventions. Being alert to the signs of distress and being equipped with a range and variety of intervention strategies and techniques provide teachers with the skills necessary to save children from impending failure.

We need to focus on success strategies in our classrooms to help all children grow to become resilient, self-directed, life-long learners: "From impossible to I'm possible."

Arthur L Costa Ed.D.
Professor Emeritus,
California State University, Sacramento

1

Classroom Intervention Strategies That Work

"The instruction we find in books is like fire.
We fetch it from our neighbors,
kindle it at home,
communicate it to others,
and it becomes the property of all."

—Voltaire

Kimmie steps over the cracked pavement and passes run-down shops next to abandoned buildings as she walks down Seventh Avenue. The store fronts have chained gates and metal bars protecting the merchants' livelihoods. She notices the bustle of inner-city traffic and smells the exhaust fumes mixed with the aromas from the corner market. Every morning she walks the eight blocks from her apartment to the local elementary school squished between two large buildings. Washington Elementary is a 60-year-old, two-story building with a dingy red brick exterior and no grass in the schoolyard. The hop-scotch lines are fading on the black top and the rusty basketball hoops have no nets. Although it feels old, Kimmie likes the stability of school and she hopes that she will be able to stay at Washington Elementary. As a six-year-old just about to turn seven, she has already moved five times in her young life. She enjoys school, yet she finds it difficult to keep pace with her classmates. She tries her best, yet she often falls behind and learns slower than her fellow first graders. Reading the books at school definitely is the hardest for her. At home she enjoys looking at the pictures and she tries to read all of the words in the one book she owns. Kimmie received the book from a nice neighbor lady who was kind enough to give it to her as a gift when she moved out of her last apartment. She looks forward to the day when she will be able to read all of the words in her book. Kimmie is a student from poverty who struggles in school.

Many of the students in our schools are a lot like Kimmie. Although these students have a variety of backgrounds, they also struggle to keep up with the expectations of school. For decades our school systems in America have worked to address the needs of all students. Many students have benefited from the efforts of our schools, and they go onto enjoy tremendous opportunities. At the same time, many of our students find traditional schooling to be difficult and overwhelming. Every day, classroom teachers, instructional coaches, and site administrators face the challenge of making sure that all students learn effectively and succeed in school.

So, How Are Our Schools Doing?

While our schools help many students, it seems that many other students fall far short of the demands of school. More and more students are falling quietly by the wayside. How should we view this situation? It is said that when tribal leaders come together to discuss the conditions and concerns of their tribe, the first question they ask is, *"How are the children?"* This inquiry is based on a belief that the tribe is only as healthy as the children. They are the barometers of success. So, how are all our children doing in school? If we look at the numbers, we will see that the overall results are less than encouraging. Over 45% of African-American and 45% of Hispanic students fail to graduate from high school with their peers (Orfield, Losen, Wald, & Swanson, 2004). Many students in our country never make it all the way through school and minority students from poverty drop out at alarming rates. Consider the following results in Figure 1.1, (adapted from Alliance for Excellent Education, 2009).

Each year 1.2 million students drop out of our schools. Statistically, those that drop out of school face higher unemployment, incarceration, and lower lifetime earnings (Sum, Khatiwada, McLaughlin, & Palma, 2008). The social

Figure 1.1 Dropout Rate Percentages in the United States

National Average	31%
Hispanic Students	45%
African-American Students	49%
Inner-City School Students	47%
Low Socioeconomic Students	50+%

and economic impact is staggering to consider. Each year more and more students are dropping out of school. Why do students drop out of school? More than 75% of school dropouts mentioned that difficulties in their ability to read in core classes contributed to their decision to drop out (Lyon, 2001). For those students who remain in school, many of them struggle to meet the demands of school. The average student from chronic poverty is more than two entire grade levels behind their peers by the time they leave middle school (Johnson, 2009). Let's take a look at another student who has difficulty keeping up with the pace of school.

> Jose is a kindergartner at Vista Nueva Elementary. He was born and grew up in the outskirts of Ciudad Juarez, and his family migrated across the river to his current home in El Paso, TX. Jose and his family speak Spanish at home, and he and his brothers and sisters speak English at school. He speaks Spanglish as he plays with his friends who live in his neighborhood. Jose avoids speaking in class, because he lacks confidence. He is unsure how to blend school conversations with his home life. Jose is a second-language learner who struggles in school.

Jose is like so many of our students who need targeted support to help them be truly effective in school. Without targeted interventions, students like Jose may find that dropping out of school seems like a better opportunity than staying in school. So, what can we do to help more students feel successful in school?

Do Our Schools Need an Intervention?

The simple answer is a resounding ***YES***—we definitely need an intervention in American education. Our schools need an intervention for the sake of so many of our children and for the collective future of our country. Interventions are usually reserved for those who are in crisis circumstances and the situation for many of our students grows more severe every day. The quality of education that our students receive has become an extremely pressing issue.

To get to the root of inequity within our country, we must improve instruction within our schools. We need to invest in educational interventions that will reach out to all of our students, particularly those who are socioeconomically disadvantaged and those that live without literate role models. It is essential that we as educators become responsive to the needs of all our students sitting in front of us every day. Our schools and our students will benefit by systematically and strategically providing effective instruction, feedback, and early intervention strategies. Allington (2010, p. 20) notes:

> We have studies involving multiple school districts and hundreds or thousands of kids demonstrating that, with quality instruction and intervention, 98 percent of all kids can be reading at grade level by the end of 1st or 2nd grade.

Our educational efforts should be directly focused on research-based instruction and classroom intervention strategies that work with our students. High-quality instruction, more responsive efforts, and targeted intervention strategies are required. Without effective classroom instruction and intervention strategies, we will struggle to reach all students. Let's take another look at a student who faces difficulty in school.

> Susie is a second grader at West Oaks Elementary. She comes to school each day with a dimpled smile that makes you want to squeeze her little cheeks. She loves cutting and pasting art projects at school, and the projects always seem to be displayed on her refrigerator at home. She is quiet and a little shy at school and she works hard to learn at school. When it comes time to read words, she struggles to grasp the sound-symbol relationship between letters and their corresponding sounds. Susie is already more than one grade level behind her classmates and getting even further behind. Susie is a struggling reader who also struggles in school.

Susie faces similar difficulties in school as Kimmie and Jose, even though their background circumstances are quite different. They each struggle academically to keep up with their grade level peers. They need intervention strategies that will help them accelerate their learning. They will need special attention to achieve grade level standards. Susie, Kimmie, and Jose, like so many other students, need classroom strategies that work to intervene and provide access and experiences on grade level.

Who Needs Classroom Intervention Strategies?

Every teacher in America will benefit from intervention strategies that meet the needs of their students. Teachers consistently are on the lookout for classroom strategies that will reach out and help even the most challenging student. For example, as we (the authors) interact with teachers and administrators at workshops across the county, the number one concern we consistently hear is, *"I need intervention strategies that I can use in my classroom to meet the needs of my struggling students."* Also, a first-year teacher wants to know, *"What do I do with my students who have a variety of problems, including socioeconomic disadvantaged status, chaos at home, and a lack of adult models?"* An elementary principal shares, *"At my school many of our students are more than one grade level behind in reading."* An experienced teacher mentions that, *"My English Language Learners often seem lost in my classroom."* While many students from a variety of circumstances may need intervention strategies, the research shows that students in three particular areas definitely need systematic classroom intervention strategies to help them develop as learners (Figure 1.2).

In the past, these low socioeconomic students, English Language Learners, and struggling readers may have been assigned to special classes. Under RTI, many of these students are now expected to succeed within the regular classroom. Let's look at some of the specific needs of these students.

Figure 1.2 Children Most in Need of Instructional Intervention Strategies (Honig, Diamond, & Gutlohn)

1. Children raised in poverty, those living in chaos, or without adult models.
2. Children who are English Language Learners
3. Children who struggle with phonological processing, memory difficulties, and speech or hearing impairments

Low Socioeconomic Students (Title I): Students from low socioeconomic backgrounds typically lack a myriad of resources that affect their results in school. Children are the largest group of individuals affected by low socioeconomic circumstances. Like Kimmie, who we introduced at the beginning of this chapter, these students need targeted intervention strategies that will help them develop their knowledge and understanding. Without early school interventions, these students face falling behind their grade level peers, dropping out of school altogether, and being at risk of a host of other unfortunate life events (Stone, Silliman, Ehren, & Appel, 2005).

English Language Learners (ELLs): Our K-2 classrooms are filling up with increasing numbers of English Language Learners. Again, these students need specific, targeted interventions to overcome the effects of learning a new language. Approximately 50% of English Language Learners drop out of school (Alliance for Excellent Education, 2009). Like so many other ELL students, Jose speaks a different language than English in the home. School is the only place where many ELL students have a chance to hear and produce language in English. These students need strategies that will help them face the challenges of learning a second language. Our ELL students need *classroom intervention strategies that work*, and they need these strategies so they can develop the language skills required to succeed in school.

Struggling Readers (Special Needs): Struggling readers come to school in many shapes and sizes. Like Susie, struggling readers find school extremely challenging. With so much learning conveyed through classroom texts and stories, difficulty with reading significantly impacts learning. It can be difficult to spot these students just by looking at them, yet get them to open a book and begin reading out loud and the issues present themselves quite readily. Research shows that unless high-need students receive targeted interventions, many will face increasing gaps in their learning and achievement (Rathvon, 2008). If no interventions are provided at an early age, then many of these students may be destined for special education support for years to come. Struggling readers need *RTI strategies that work*, and they need them early and often.

Low socioeconomic students, English Language Learners, and struggling readers need consistent classroom interventions.

An Instructional Toolbox of Interventions

The metaphor of the instructional toolbox fits very well with classroom intervention strategies. One of the most important things for a teacher to do is continually add to their instructional toolbox. Irvin, Meltzer, & Dukes (2007, p. 76) point out, "Interventions provide students with the tools and strategies they need to make great strides in literacy development."

The research shows that many teachers get by with a very small instructional toolbox of strategies. The average teacher only uses four to five strategies effectively and efficiently with students. While these strategies may go a long ways, students will benefit from a variety of additional strategies that will reinforce their intervention needs. Throughout this book you will learn 25 strategies that are specifically targeted for K-2 classrooms. When we add to our intervention toolbox, then our instructional repertoire will expand its capacity to fix the issues that face so many of our students.

What Does a Response to Intervention (RTI) Model Look Like?

Let's take some time to see what a Response to Intervention (RTI) model of instruction looks like. The RTI model has become a growing movement across the country. RTI has become the national model for addressing struggling students' needs. Many states like New York and Iowa have already passed legislation that makes RTI the official educational response for their schools.

The RTI process typically follows a Three-Tier (Tier I, Tier II, and Tier III) model that begins with a focus on interventions in the classroom (Figure 1.3).

Figure 1.3 **Three-Tiered Response to Intervention (RTI) Model**

Tier III: Intensive Interventions
(5% of students)

Tier II: Targeted Interventions
(15% of students)

Tier I: Core Classroom
Instruction Interventions
(80% of students)

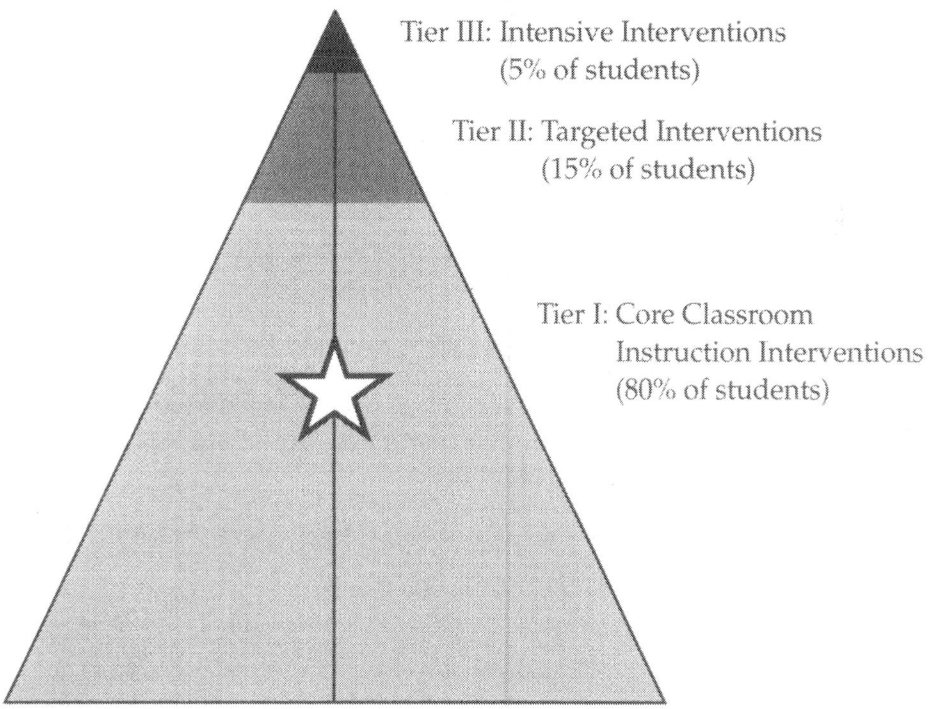

Figure 1.4 **RTI Matrix**

RTI Tier	Intervention Level	Targeted Students	Best Practices	Expected Results
Tier I	Classroom Intervention	All Students	♦ Universal Screening ♦ Research-based practices ♦ Explicit Instruction ♦ Consistent Prog-ress Monitoring	80% of students at grade level
Tier II	Targeted Intervention	Some Students	♦ More Time ♦ More Attention ♦ Increased Support ♦ Targeted Progress Monitoring	15% of students at grade level
Tier III	Intensive Intervention	A Few Students	♦ Individualized instruction ♦ Individualized assessment based on student needs	5% of students at grade level

Effective classroom interventions are designed to meet the needs of 80% of the students. For students who need additional strategic interventions, more time should spent in targeted small-group instruction (4 to 6 students), which should meet the needs of an additional 15% of students. For students who still struggle to meet grade-level standards, they should receive intensive interventions. A few students may need interventions that provide intensive support to reach the remaining 5%. In Figure 1.3, the three colored sections identify three different levels of interventions:

Tier I – **Classroom Interventions** are designed for **All Students,** and they should reach approximately **80% of students**.

Tier II – **Strategic Interventions** should help **Some Students,** an additional **15% of students**.

Tier III – **Intensive Interventions** should target **A Few Students,** the final **5% of students**.

The most important interventions that reach the greatest number of students are high-quality classroom interventions (Figure 1.4). For students who continue to struggle, they may be given targeted interventions that target specific student needs. Finally, students who continue to struggle should be provided intensive interventions.

Beginning an RTI model should start with universal screening of all students. A variety of quick and simple diagnostic assessments can be used to universally screen students. Most of the major curriculum publishers now provide universal screening assessments to determine which students may need additional help to meet grade-level standards. Classroom teachers should also check for understanding and monitor student progress to get continuing feedback so that a student's academic needs can be addressed.

The Key to K-2 Classroom Intervention Strategies

The foundation for an effective intervention program starts with strategies that are implemented by classroom teachers. When we implement intervention strategies that work, we can have the confidence our students will reach success. Mellard and Johnson (2008, p. 70) note:

> Most students will achieve academic success when provided with high-quality Tier I instruction. As such, Tier I can help reduce the incidence of so-called instructional casualties by ensuring that students receive appropriate instruction accompanied by progress monitoring.

Other research books on RTI have focused on the intensive assessments and individualized methods focused on the third tier of RTI for educational specialists (special education, etc.). This book, *RTI Strategies that Work in the K-2 Classroom*, is especially designed for Tier I whole-group and small-group instruction. It provides everyday K-2 classroom teachers with effective intervention strategies for low socioeconomic students, second-language learners, and struggling readers. Mellard and Johnson (2008, p. 70) go onto add the following insights regarding effective Tier I interventions provided by the classroom teacher.

> Tier I is particularly important for two reasons. First, it represents the first gate in a system designed to accommodate the diverse learning needs of all students. Thus, Tier I provides the foundation for instruction on which all supplementary interventions are formulated. Second, since Tier I focuses on all students, it is the most cost-effective means of addressing the population of learners.

Figure 1.5 **Characteristics of Effective Interventions**

Struggling Students Benefit from Classroom Intervention Strategies that

- include explicit, well-organized (systematic) instruction as well as opportunities to read connected text;
- are provided in small-group or one-on-one formats;
- provide for 20–40 minutes at least three to five times per week;
- provide extended opportunities for practice, including guided, independent, and cumulative practice with teacher feedback;
- are provided in addition to regular classroom reading instruction; and
- include continuous progress monitoring.

The accumulated research on interventions for young students in the early grades has found that effective interventions achieve a variety of objectives. Take a look at Figure 1.5 (adapted from Glover & Vaughn, 2010) regarding classroom intervention strategies.

K-2 teachers who provide a variety of quality whole-group and small-group instructional strategies will see the vast majority of students meet grade-level standards and objectives.

Implementing Classroom Intervention Strategies

When schools consider adopting an RTI model, they typically focus on placing students in regular classrooms instead of sending them off to portables that specialize in struggling students (Title I, ELL, Special Ed, etc.). This is potentially a very positive opportunity if the regular classroom teacher also receives additional strategies to effectively meet the additional instructional needs of these struggling students. With this influx of struggling students, we need classroom intervention strategies that can be effectively implemented, so that all students can achieve grade level standards with appropriate support. As noted earlier, K-2 classroom teachers are clamoring for intervention strategies that will effectively reach all students. Effective instruction and intervention strategies are at the heart of Tier I whole-group instruction and Tier II small-group instruction. Let's look at the definition of strategy from Dictionary.com (2010):

Strategy

1. A plan, method, or series of maneuvers or stratagems for obtaining a specific goal or result
2. Skillful use of a stratagem.

The classroom intervention strategies covered in Chapters 2 through 6 will provide K-2 teachers with research-based methods that will enhance Tier I and Tier II interventions. The research emphasizes that high-quality whole-group instruction and targeted small-group strategies provide the most effective way to build a RTI program with a firm foundation (Wixson, Lipson, Scanlon, & Anderson, 2010). The strategies in this book have been designed to help classroom teachers substantially strengthen their whole-group instruction and small-group interventions for struggling students. Our students will find that these strategies will help their struggling students access grade-level text, engage their interests, structure their learning, and create meaning from K-2 standards.

Strategic Arrows that Hit the Target

In addition to the metaphor of the toolbox, another metaphor that helps convey classroom intervention strategies is that of the archer with his quiver and arrows. As teachers, each one of us should have our own instructional quiver of strategic arrows. We can use these intervention arrows to target in on our students' learning needs. The intervention arrows are intervention strategies that we use to hit the heart of student learning. These arrows help us target in on our students' intervention needs and point them in the right direction. The more intervention arrows or strategies we have available, the better we will be able to use these strategies to target in on the issues that face our struggling learners. So how many arrows do you have in your intervention quiver? This book provides *25 instructional strategies* to target in on our students' educational concerns. These arrows/strategies will help you hit the mark for your students' learning needs. As you learn how to use these intervention arrows in your quiver, you will be able to hit the target for the specific needs of your students. This will take a little practice, yet in no time at all you will be consistently be hitting the mark. Using interventions effectively is much like being a skilled archer/marksman. With the wide range of issues that children bring with them to school, every teacher needs to add strategies to their intervention quiver or instructional toolbox.

Effective Intervention Grouping

Targeted classroom interventions should be started early and they should be used often. Students need intervention supports provided to them the same day that they struggle to grasp objectives and skills. Each day teachers should devote time working with a small group of four to six students who need additional support mastering the skills and objectives outlined in the daily lesson. Allington (2010) encourages at least a half hour a day of classroom interventions for students who need additional time and attention.

> In 1st grade, most of the studies have recommended either a half-hour or 45 minutes a day, five days a week, usually for a period of roughly 20 weeks, as an initial shot at it. At that point, some kids still may not be up to grade level. But if you give them another 20 weeks, you can be down to 2 percent of kids who aren't reading at grade level.

Teachers should monitor progress during the lesson time and frequently check for understanding (Fisher & Frey, 2007). The sooner we begin interventions, the sooner we can get our students back on track. We should provide interventions in kindergarten (Rosenfield & Berninger, 2009). In fact, the earlier students receive intervention strategies, the sooner they will be able to produce results. A quality RTI program begins in the classroom. The classroom teacher is best positioned to see concerns and address them right away. Let's look at three areas of focus that the classroom teacher needs to master as they provide effective intervention strategies:

Whole-Class Intervention: Interventions at this level are provided by classroom teachers to support the core instruction for all students and are designed to meet the needs of at least 80% of the students. Effective teachers recognize where struggling students are falling through cracks and gaps in their learning and they provide classroom intervention strategies to meet the needs of their students.

Targeted Small-Group Intervention: The teacher will invite a group of four to six students to the small-group instruction (peapod) table to receive more time and attention on intervention strategies that will help students achieve necessary knowledge, skills, and objectives. Targeted small-group instruction typically requires teachers to devote 30 minutes to ensure that struggling students have sufficiently learned the daily objectives. (The next section in this chapter will discuss core components of classroom intervention strategies that work)

Independent Practice: While the teacher is focused on small-group intervention, the rest of the students in the class complete daily activities that allow students to practice the skills taught in the core lesson. Teachers can provide students with activities in two categories:

- **Must Dos** (Activities selected by the teacher that reinforce the lesson and provide students with independent practice.)
- **May Dos** (Activities selected by the teacher to extend and enhance learning.

The teacher writes on the board the daily Must Dos and May Dos that will help the bulk of the class (80%) demonstrate their mastery of the grade level skills and objectives. Typically a teacher assigns one or two Must Dos and provides one or two May Dos that students can work on independently so teachers can use this time to concentrate more time and attention on small-group instruction for students who struggle.

Always ask yourself *"Is this child performing within the range of grade level expectations? If not, which strategies will make a difference?"* Every classroom needs quality whole-group instruction and small-group interventions within the regular classroom. Our students need intervention strategies each and every day to help them accelerate their learning. Small-group interventions provide students with two important resources:

- **More Attention** (Small-group attention focused on strategies)
- **More Time** (Additional time with students working on strategies)

As struggling students receive more attention and more time, then more students will be able to achieve K-2 grade-level expectations. Highly successful schools and highly successful teachers provide their students with targeted intervention strategies that help each student achieve classroom success.

Core Competencies for Classroom Intervention Strategies that Work

Whether we are an administrator, teacher, or instructional coach, we first need to develop our own understanding of intervention strategies that work for students. We need in our instructional toolbox a repertoire of strategies that will achieve effective results. This instruction needs to have built-in interventions that address the varying needs of students. Our students need specific intervention strategies that will develop their ability in reading, writing, numeracy, listening, and speaking.

Core Intervention Competencies

- **Listening** (strategies that help students focus and understand better)
- **Reading** (strategies that help students comprehend and handle text)
- **Numeracy** (strategies that help students compute and solve equations)
- **Speaking** (strategies that help students dialogue and engage with peers)
- **Writing** (strategies that help students compose and construct ideas)

Our instructional arrows must hit the mark for the desired outcomes. And we need a variety of intervention strategies in our instructional quiver so that we can hit the mark and meet the different needs of our students.

What Do Classroom Intervention Strategies that Work Look Like?

The primary challenge is that a response to intervention is only a model or framework for restructuring our approach to students. In many ways, an RTI approach restructures how students receive support. RTI looks to begin interventions much earlier for students who need help and it emphasizes the classroom as the primary place interventions occur. The top challenge right now in education is classroom teachers lack many of the targeted intervention strategies that will produce the desired results for struggling students. Targeted intervention strategies are used by effective teachers to meet the instructional needs of students. While there are many things that meet the many needs of students, the following four areas are crucial to designing effective interventions (Figure 1.6).

As we provide interventions that are specifically designed to support access to learning, student engagement, structured activities, and making meaning, then our students will increase their academic achievement. Let's look at each of these four keys to successful interventions. Effective Classroom Intervention Strategies provide students with the essential support needed for students to be successful academically in school.

Figure 1.6 **Characteristics of High-Quality Intervention Strategies**

1. **Access** (Language Support)
2. **Engagement** (Interactive and Internal Processing)
3. **Structure** (External and Internal)
4. **Meaning** (Pulling it all together)

Providing Access (Language Support)

The number one reason that the vast majority of our students are unable to succeed in school is their difficulty accessing the knowledge provided at school. Access is difficult for many students, because students lack the academic language and background knowledge needed to successfully access the learning provided in teachers' instruction and classroom textbooks. Brown-Chidsey, Bronaugh, and McGraw (2007, p. 16) note it is essential that "each child has access to quality instruction and that struggling students—including those with learning disabilities—are identified early and receive the necessary support to be successful."

Accessing learning through language is essential for success. Language provides all of us the ability to access learning. Students need access to learning through high quality interventions that support language and learning (Shores & Chester, 2008). At the same time, a lack of language can limit one's ability to learn in school. Ed Hirsch (2003, p. 22) notes, "It is now well accepted that the chief cause of the achievement gap between socio-economic groups is a language gap."

The most important type of language at school is academic language. While it typically takes a second language learner one to two years to learn the casual language needed to talk in the neighborhood, on the playground, or with a classroom peer, it takes much longer for students to learn academic language (Figure 1.7). The reality is that many students never learn the academic language needed to succeed in school (Johnson, 2009).

Figure 1.7 **Academic Language**

Language Register	Number of Years Needed to Learn
Casual Language	1 to 2 years
Academic Language	5 to 7 years

Even though second-language learners can learn the academic language in five to seven years, without targeted interventions many students never learn academic language and they struggle throughout school. Increasing student access to learning through strategies focused on language builds student progress (Maanum, 2009). Now let's look over the a framework for developing reading from John Shefelbine outlined in the Reading/Language Arts Framework for California Public Schools (California Department of Education, 2007) (Figure 1.8). Note how academic language is essential in helping students transition from decoding (learning to read) to comprehension (reading to learn).

Figure 1.8 **Framework for Reading**

DECODING Learning to Read					COMPREHENSION Reading to Learn				
Word Recognition Strategies				Fluency	Academic Language		Comprehension Strategies		
Concepts of Print	Phonemic Awareness	Phonics	Sight Words	Automaticity	Background Knowledge	Brick and Mortar Vocabulary	Syntax & Text Structure	Comprehension Monitoring	(Re)organizing text

Adapted from John Shelfbine/Developmental Studies Center

Students from poverty, English Language Learners, and struggling readers all have difficulty accessing learning because they lack the background knowledge and essential language that leads to effective learning.

Providing Engagement (Interactive and Internal Processing)

Students need interventions that are engaging and use all of the senses. Effective interventions are kinesthetically engaging, they are auditorally resonating, and they are visually stimulating. There is a direct correlation between engagement and academic achievement, particularly for students from poverty (Wilms, 2003). The most powerful models of instruction are interactive. Instruction actively engages the learner, and is generative. Instruction encourages the learner to construct and produce knowledge in meaningful ways. Students teach others interactively and interact generatively with their teacher and peers. This allows for co-construction of knowledge, which promotes engaged learning that is problem-, project-, and goal-based. Ivey and Fisher (2006, p. 81) state,

> It is easy to spot engaging instruction within an intervention. Students are eager to read and write.

Some common strategies included in engaged learning models of instruction are individual and group summarizing, means of exploring multiple perspectives, techniques for building upon prior knowledge, and so on. Great intervention strategies are:

Engaging (Instruction and interventions that are engaging help students find excitement in learning. They become intrinsically motivated.)

Interactive (Students learn to work collaboratively and connect with their fellow classmates as they learn.)

Energizing (Effective strategies build energy and help students maintain their focus as they accelerate their learning.)

In addition to engaging the external senses, effective interventions also engage internal learning processes.

Most importantly, our students need to be internally engaged in extending, expanding, and elaborating on their learning. Reflective practices and monitoring progress help students engage internally and identify learning successes and areas for additional attention.

Providing Structure (External and Internal)

Intervention strategies need to scaffold or provide workable chunks that students can grasp as they learn processes. Explicit instruction means that the students can see the specific steps needed to accomplish a task. Riccomini and Witzel (2009, p. 39) emphasize,

> The organization of RTI rests on a tiered system to structure instruction and interventions based on student needs.

Students need frequent feedback along the way to make sure that they are on track. Students need learning processes modeled for them. Hawley, Rosenholtz, Goodstein, and Hasselbring (1984, p. 127) write,

> What teachers do in the structuring of learning opportunities and the provision of instruction is at the heart of the contribution schools make to the academic achievement of students.

Particularly the internal processes that signify great learning like summarizing, understanding, and making connections provide the structures that strengthen and build learning. These processes and many others need to be clearly outlined in specific instructional steps that will help students learn and comprehend knew information. You will notice as we jump into the five upcoming chapters outlining the key strategies that specific steps are provided for each strategy.

Providing Meaning (Pulling It All Together)

The ultimate objective of any lesson is that students make meaning of the information they are learning. Many students go through lessons in class without really developing any meaningful learning. Consider the follow insight from Fullan (2001, p. 46):

> Acquiring meaning, of course, is an individual act but its real value for student learning is when shared meaning is achieved across a group of people working in concert.

When students lack the academic language and academic literacy, quality classroom interventions can help students negotiate with the text, with their peers, and with their own sense of meaning. Bearne, Dombey, and Grainger (2003, p. 158) state,

> Negotiating meaning and making sense are basic to all literate behavior.

It is through the processes of negotiation with others that we develop greater literacy skills and gain insight into new learning experiences. Learning is a process of exchange where our students negotiate with their own understanding so they are able to make meaning. Interventions should encourage students to negotiate meaning with their teacher, classmates, and the author of classroom texts.

Intervention Strategies that Work

We have many students who need us to respond to their intervention needs. In the next five chapters we will look at classroom strategies that will support our students' Tier I, Tier II, whole-group, and small-group instructional needs. Every teacher can add to their toolbox of intervention strategies. Hoy and Miskell (2002, p. 134) found that teachers typically use only a handful of instructional strategies:

> Moreover, ecological research in classrooms repeatedly finds that teachers rely on just three or four instructional routines to accomplish the majority of their instructional work.

We need to add many more intervention strategies and routines to our instructional repertoire, so that we can meet the needs of every student. In Chapters 2 through 6, we will spell out twenty-five intervention strategies that will support Tier I and Tier II RTI instruction with the whole classroom or small groups of students.

Summing It Up

Because our schools are filling up with more and more struggling students, we need more intervention strategies that work in every classroom. Students like Kimmie, Jose, and Susie need quality classroom instruction with effective intervention strategies. English language learners (ELL), low socioeconomic students (Title I), and struggling readers will benefit from interventions that specifically target their needs. Interventions should be provided early and often to students as soon as they show they are struggling with grade-level objectives and skills. The book focuses on strategies that will help the classroom teacher provide high-quality Tier I and Tier II interventions. Rather than focusing on ways to restructure the external dynamics of our schools, we should use RTI to focus on better ways to instruct and provide intervention strategies for our students. Struggling students need *RTI Strategies that Work* so they can keep up with grade-level expectations and learn at the same rate as their peers. Effective classroom intervention strategies help students with needed **access**, **engagement**, **structure**, and **meaning**. While the strategies in this book will benefit all students, struggling students will enjoy significant benefits from targeted small-group strategies that provide more time and more attention. The next five chapters will explain how these intervention strategies can support all of our students. As we fill up our intervention toolbox with more strategies we will have the ability to meet the demands of all students. The last chapter of the book outlines how we can make continuing progress as we discuss student achievement in Professional Learning Communities (PLCs). Adding these strategies to our instructional arsenal will help us target student's instructional needs and change the trajectory of their future.

Reflection

1. How do you provide classroom interventions to your K-2 students who need RTI support?
2. What classroom intervention strategies do you use on a daily basis to improve reading, listening, numeracy, speaking, and writing?
3. What was the last intervention strategy that you added to your instructional toolbox or classroom intervention quiver?

2

K-2 Intervention Listening Strategies

"To listen well is as powerful a means of influence as to talk well, and is essential to all true conversation."

—Chinese Proverb

Billy dresses himself every morning. However, he sometimes looks like he slept in his clothes. His shoes are smaller than his feet and he takes them off in the morning only to put them back on for recess. Billy never seems to get his clothes to match the weather—too often he is cold for lack of a coat. He always wears long-sleeved shirts and unconsciously he pulls his clothes around his small body. There are days that are good for Billy and days that are clearly bad. His teacher does not understand what is wrong but she knows that Billy is "unparented." No one comes for parent conferences, no one returns calls from the school, and he never takes his work home. Billy does not make eye contact with most adults or peers and he has difficulty listening in school. Billy is disconnected and lives in tough circumstances. He is a high-needs child who is a struggling student.

As we begin each of the upcoming strategy chapters (2 through 6), we will look more deeply into the essential components of high-quality intervention strategies. We will consider what the research says about our students and the strategies they need. Before we jump in, let's take another look at the students who are most in need of instructional interventions (Figure 2.1 on page 22).

Figure 2.1 **Children Most in Need of Instructional Intervention Strategies (Core, 2001)**

1. Children raised in poverty
2. Children who are English Language Learners
3. Children who struggle with phonological processing, memory difficulties, and speech or hearing impairments

Why Do So Many Students Need Interventions?

It seems that the number of kids like Billy seems to be increasing in our classrooms. They have difficulty paying attention in class and they often miss many days of school. The research shows that many of the students like Billy enter kindergarten already academically behind. These students struggle to access essential learning objectives because they lack many of the language skills they need. Hart and Risley (2003) found through their analysis of communication patterns in welfare, working class, and professional homes that children are profoundly affected by the conversations between parents and their children. The research revealed that kids from welfare homes only had working vocabularies that were half the size of their wealthier peers. Since a student's vocabulary is highly correlated with his or her knowledge and ability to learn, this difference in vocabulary size due to socioeconomic factors has a huge impact on a student's educational future. As the chart below shows (from Hart & Risley, 2003), this language gap shows up as early as age three, before children even enter school (Figure 2.2).

Figure 2.2 **Number of Words in Children's Vocabulary**

	By Age 3	By Age 7
Children from welfare families	500 words	5,000 words
Children from working class families	700 words	7,000 words
Children from professional families	1,100 words	11,000 words

Hart and Risley were astounded by the results of their research and they searched further to determine the key factor that played a role in causing this dramatic difference in children's vocabulary. After recording and

Figure 2.3 The 30-Million-Word Gap

	Words heard per hour	Words heard in a 100-hour week	Words heard in a 5,200-hour year	Words heard over a 4-year period
Welfare	616	62,000	3 million	13 million
Working Class	1,251	125,000	6 million	26 million
Professional	2,153	215,000	11 million	45 million

listening to countless hours of communications between adults and the children, they discovered that welfare parents only spoke and provided one third the amount of words to their children in the home compared to their more wealthy peers. Over the course of a four-year period, this means that students from poverty will hear 13 million words from the adults around them, which is on average 30 million fewer words than students with professional parents. Much of this has to do with the types of conversations that happen around the dinner table where professional parents ask their kids about things like "What did you learn at school today?" and "Did you make any new friends?" Let's take another look at Hart and Risley's (2003) analysis of language and learning (Figure 2.3).

So the home environment plays a huge role in how much language a student hears from a parent or parents. We know that socioeconomic background is a huge factor in determining educational outcomes for kids. The reason socioeconomic issues impact education so dynamically is the direct correlation to language development. The research shows that poor students receive significantly less language input from parents (one third compared to wealthier peers) and this results in significantly less language output by students (one half compared to wealthier peers). Unless schools specifically address this language gap between poor kids and their wealthier peers with targeted intervention strategies, then the gap just grows over time.

Three times (3x) the listening input provided to young children roughly equates to two times (2x) the vocabulary knowledge output.

By the time students enter twelfth grade, wealthy students continue to have a vocabulary that is twice as large as their poorer peers (Johnson, 2009). Schools and classrooms are the only place where this discrepancy can really be successfully addressed. Marzano (2001) has shown that high-quality classroom instruction and key intervention strategies can address the language gap and help all students perform at grade level.

How Else Does Language Impact Learning?

One more thing needs to be said about the research comparing students from welfare, working class, and professional homes. It was also discovered that in addition to the quantity of language heard by children in the home, there was also a dramatic difference in the quality of language heard in the home (Figure 2.4). For example, welfare class parents typically give only half as many positive affirmations like "Good job" compared to negative discouragements like "Knock it off" or "You're wrong." At the same time they noticed that working class parents provided twice as many affirmations like "Way to go" and "That's terrific" compared to discouragements like "Stop that." And most surprising, professional class parents provided six times as many positive affirmations like "Nice work," "Great effort," "Keep sticking with it," "Wonderful job," "Fantastic," and "That's a good idea." compared to discouragements like "No, you may not."

It is important to note that a poverty of socioeconomic background affects language foundations and a poverty of language is highly correlated to a need for instructional interventions. The tremendous difference between the quantity of language vocabulary developed (high socioeconomic students have twice the amount of vocabulary than their low socioeconomic peers when they enter school) and the quality of language communicated (high

Figure 2.4 **Average Difference in Positive/Negative Communication at Home**

Children from Welfare Families	26,000 affirmations vs. 57,000 discouragements (1–2 ratio)
Children from Working Class Families	62,000 affirmations vs. 36,000 discouragements (2–1 ratio)
Children from Professional Families	166,000 affirmations and 26,000 discouragements (6–1 ratio)

socioeconomic students hear over six times the amount of affirmations than their lower socioeconomic peers at home) is an astounding gap. So, the language gap is a primary factor behind the achievement gap and intervention strategies need to address the gaps in language and literacy. Students from poverty need *classroom intervention strategies that work* to fill the gaps that exist in their language, literacy, and learning.

Developing Good Listeners

Children need instruction that includes a wide repertoire of instructional strategies to help them really "learn to listen to words" and understand what is being said by their teachers. Teaching listening has become a major concern because many students have limited time being read to by adults and have limited positive dialogue with adults; rather, these students have spent extensive amounts of time in front of televisions and computer screens and in chaotic, loud circumstances. Jim Rose (2006, p. 3) notes:

> The indications are that far more attention needs to be given, right from the start, to promoting speaking and listening skills to make sure that children build a good stock of words, learn to listen attentively and speaking clearly and confidently. Listening and speaking, along with reading and writing are prime communication skills that are central to children's intellectual, social and emotional development.

Listening to words entails many discrete skills for the children. These listening prompts can be presented in a four-step process:

Listening Prompts

1. Listen to the speaker with both your ears and eyes.
2. Watch what the speaker does to help you hear and understand the words.
3. Take a deep breath and sit with your hands in your lap.
4. Choose to speak only when your hand is raised and the teacher calls on you.

Developing intervention strategies for listening adds to our teaching repertoire and we strengthen our ability to target in on students' instructional needs. As we get ready to jump into researched intervention strategies that work, you will notice that each strategy follows the same easy-to-follow pattern.

Intervention Strategy Format

- **Introduction to the strategy** (Overview and research)
- ***What* the Strategy Looks Like** (Key components of the strategy)
- ***How* the Strategy Works** (Step-by-step instructions)
- **Kindergarten and English Language Learner** (Scaffolding for nonreaders)
- ***Why* the Strategy Works** (Reasons the strategy reaches struggling students)
- **Progress Monitoring** (Three methods for monitoring student progress)

Throughout the book, each of the 25 strategies will have charts, graphs, sample assignments, or pictures that will help exemplify and model the strategies. All of these resources will help your students grab a hold of the strategies and really help them learn. Before we get into the upcoming strategies, we should emphasize what it will take to succeed with these strategies, so that they really work.

> **Our students need to engage in an intervention strategy at least 6 to 12 times before they will become proficient at the strategy.**

It is important that we weave intervention strategies into our current learning objectives to make learning more accessible, more engaging, more structured, and more meaningful for our students. So with all of that said, let's jump into Listening Intervention #1.

Listening Intervention #1: Active Listening

As students enter school there is so much for them to learn. This new information is best learned when young students develop actively listening. It is often assumed that the newest students come to school fully prepared to LISTEN TO THE TEACHER. We often take listening strategies for granted, because it is assumed that our students come to school as capable listeners. Students who have difficulty learning typically have never been explicitly

Figure 2.5 **Average Amount of Time in Class Spent in Different Literacy Areas:**

- ♦ 50% of time spent on Listening
- ♦ 25% of time spent on Speaking
- ♦ 15% of time spent on Reading
- ♦ 10% of time spent on Writing

taught to listen. Children in chaotic circumstances may be loud and disruptive or lost and disengaged, because they lack skills to listen effectively. Listening plays a vital role in learning for students who may appear difficult, defiant, or distressed. It is important that students focus on their listening skills to ensure that they are prepared for the more rigorous challenges of comprehension and attaching meaning to what they hear and see. Students will spend the largest part of their learning time at school on listening as noted in the research from Rubin (1994). An unsuccessful listener can quickly become a struggling learner (Figure 2.5).

Students need to become active listeners who focus on the purpose, objectives, and content of learning. They need to know how to transform what they hear into topics for discussion and writing. Listening comprehension precedes reading comprehension (Cain & Oakhill, 2007), thus, we can see that good listeners become good readers.

What Active Listening Looks Like

As teachers, we have a clear idea in our minds of the purposes for our communications to students. Young students may be unaware that when they listen, they should focus on the variety of purposes for communication. Explicitly outlining listening purposes for students will help them to provide their attention more actively to the message being conveyed in class. The practice of activating prior knowledge helps students focus on the purpose, and it kick starts young student's motivation (Dolezal, Welsh, Pressley, & Vincent, 2003). Good listeners make frequent predictions in their minds about what will happen next. Remember, good speakers communicate in whole sentences with proper emphasis and ask students to do the same. The following lists of active listening will make great classroom posters to remind students (Figure 2.6 on page 28).

Figure 2.6 What Good Young Listeners Do (Ask students to...)

- **Identify the listening purposes** for a lesson or activity.
- **Activate prior knowledge** by connecting the lesson objectives and content to what they already know.
- **Make predictions** about what they will hear next in a story or lesson.
- **Ask clarifying questions** to check listening and understanding of the main ideas.

We can ask our students the following questions to help them activate their prior knowledge (Figure 2.7).

Figure 2.7 Questions for Activating Prior Knowledge (Ask students...)

- What do you already **know about this topic**?
- What are **some experiences** you have had with this concept?
- Where or when have you **heard about this before**?
- What **words describe** what you think of about this topic?
- What are other **words that connect** to this concept?

We can ask students to make the following predictions to help them anticipate the learning that will be coming next (Figure 2.8).

Figure 2.8 Helping Young Students Make Predictions (Ask students to...)

- Make a prediction about what will happen next in **time order**.
- Make a prediction about what will happen next as a **cause and effect**.
- Make a prediction about additional **descriptive details.**
- Make a prediction about how the information can be **compared or contrasted.**
- Make a prediction about the **final result.**

The question frames below can help clarify students' understanding of the main points they should understand from listening (Figure 2.9).

Figure 2.9 **Asking Clarifying Questions (Ask students the following questions...)**

- **Today's lesson** is about...?
- The **main idea** is...?
- This is **important** because...?
- Another **key idea** is...?
- This also **matters** because...?
- In **summary**, today's lesson is about...

As we clarify the listening purposes for our students, they will be better able to focus in on learning objectives (Figure 2.10).

Figure 2.10 **Outlining Five Types of Listening Purposes (Explain to students about...)**

1. Listening **to understand and follow instructions**.

   i.e., Please write your name on the top of the paper before we begin to write.
2. Listening **to receive information and remember**.

   i.e., It is important to learn how to hear all 43 sounds in the English language, and then pronounce these sounds properly.
3. Listening **to seek enjoyment and be entertained**.

   i.e., In the story of the ant and the grasshopper we learn that it is important to be prepared.
4. Listening **to empathize and give support**.

   i.e., After listening to Stephanie's story about her turtle dying, what can we do to help her deal with this loss of her pet?
5. Listening **to examine and evaluate**.

   i.e., Do you believe that all three of the pigs made a good decision, and what would you do if you had to build a house?

Throughout the book we will present brief scaffolds that adjust the strategies for students who are kindergartners or English Language Learners who are at the earliest stage of reading. These tips will help structure the intervention strategies for these young students. Young students benefit from a poster that provides pictures of key listening purposes and reminds them to focus on these purposes (Figure 2.11).

Figure 2.11 **Listening Chart for Students**

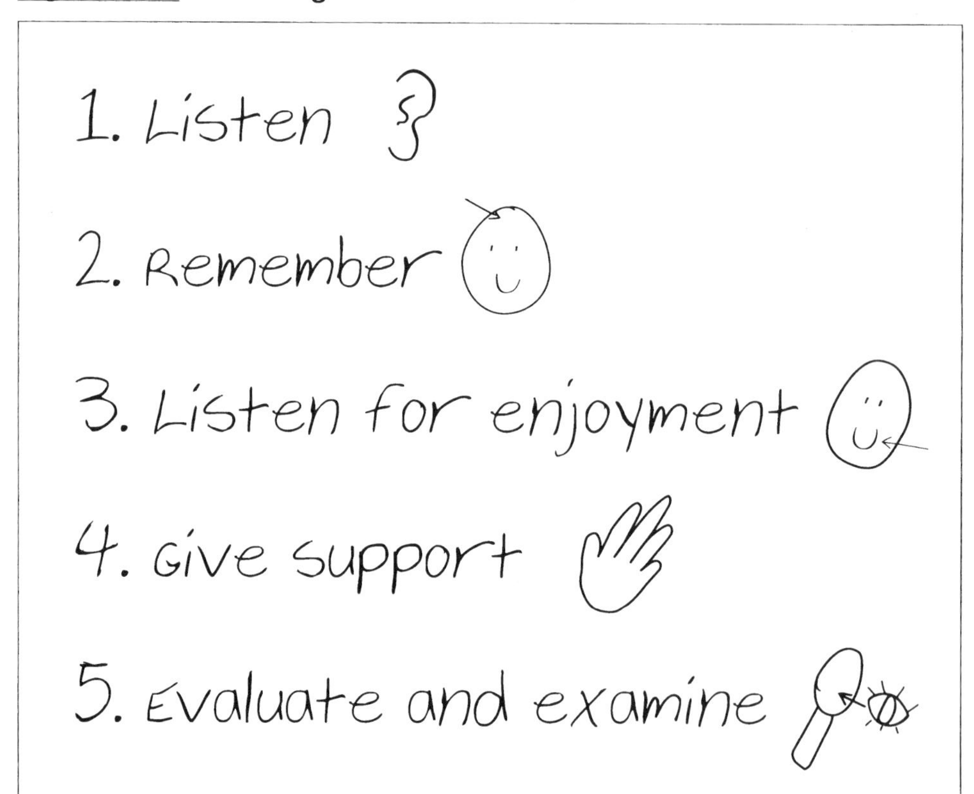

K and ELL Scaffold:

Provide kindergartners and students who have limited language very short segments of information (just a few minutes at a time) before frequently checking for understanding.

How Active Listening Works

This strategy encourages students to be aware of how they focus their attention. Young students often have little insight into their literacy skills as listeners. The following assignment can help your students determine listening purposes.

1. Write an **objective on the board** and clearly outline for students the purpose of the listening focus.
2. **Ask students to think of prior knowledge** that relates to the lesson objective or topic.
3. **Ask students to predict** what they will learn.
4. List on the board the **five types of listening purposes** (see Instructions below).
5. **Share out loud a statement** or request (i.e., open your books to page 92).
6. Ask students to figure **out which listening purpose category** this statement should be placed (See Assignment on page 32).
7. **Explain to a neighbor** their choice and justify why.
8. **Provide multiple purposes** and share answers.

Instructions **Write the five types of listening purposes on the board.**

1 = Understand and follow instructions
2 = Receive information and remember
3 = Seek enjoyment and be entertained
4 = Empathize and give support
5 = Examine and evaluate

Why Active Listening Works

Listening is the first phase of communicating and is imperative for developing rapport, inquiry, and discerning positive reinforcement. As our young students become better listeners, they will become better learners. Many students may have received less input and have smaller vocabularies (welfare students receive half the amount of wealthier peers), so we need to be explicit about how we expect them to listen and learn.

Assignment Active Listening Purposes

1. Students, please get out your math books and turn to page number 46.
2. Gather around the rug as I read you the story of "Goldilocks and the Three Bears."
3. What do you believe will be the result if we add soda to vinegar?
4. Julie just lost her pet turtle. What can we do for her to help her feel better?
5. Whenever we add an "e" following a consonant it usually produces a long A sound.
6. We will exit the bus and line up single file with your groups before we enter the museum.
7. Johnny is going to share a funny story that happened to him this week.
8. Look over your neighbor's paper and determine if they followed all of the steps for the assignment.
9. When adding numbers with multiple digits together you must line the numbers up properly.
10. Bobby broke his leg. What can we do to help him get his lunch at lunchtime?

Active Listening Progress Monitoring

Effective teachers actively monitor students' ability to listen effectively. Young students often have a limited attention span, so they need:

- To be observed sitting up squarely and to make sure that they maintain eye contact with whomever is speaking.
- A clear lesson objective and the ability to connect the lesson content with prior knowledge.
- Frequent check-ins for understanding (every couple of minutes). The younger the students, the more frequent their understanding should be checked to make sure they are getting the gist.

Some of the ways that a teacher can check for students' ability to access learning through listening is by considering the following questions: Do you have "blurters" disrupting the flow of a lesson? Are there problems with understanding the directions given for a lesson? Can you see the children listening: eyes on you, leaning forward, nodding or making facial gestures acknowledging what is being said? Will all students be able to sequence what you said, use the language you use, or translate what you are talking about into their prior experience of background knowledge? A "yes" to any of these questions should cause personal reflection of how instruction is delivered and should require an identification of the students that exhibit this classroom behavior.

Listening Intervention #2: Word Cards

It is necessary to help students shift from hearing words with their ears to seeing words with their eyes. Words are represented by symbols or letters, and students who come from low socioeconomic environments often have limited language and life experiences. These students often have difficulty creating pictures for the words they hear or read. For example, ask your students to share with you what they see when you say the word "CAT." (Take a moment and look about the word "CAT" and think about what you see).

CAT

It is interesting what responses you will receive from students as you ask them to consider the word "CAT." Some will see an orange, fluffy tabby. Some will see a black cat. Others may see a kitten or a dirty alley cat. On the other hand, struggling learners may only see letters or nothing at all. Good listeners see vivid, colorful pictures when they hear or read words. Making a connection between the words they hear and vivid pictures is an important skill for struggling students to develop. The Word Card strategy helps students connect words to vibrant pictures. Brisk and Harrington (2000, p. 42) note that the Word Card strategy is important for developing young language learners.

> The Word Card approach helps the students understand the sound-symbol connection of words. For bilingual students, the use of Word Cards can also help increase and enhance vocabulary and language use in both the first and second language.

Students need plenty of opportunities to be introduced to new words and envision pictures that represent these concepts.

What the Word Cards Intervention Looks Like

The word cards provide the students with the necessary tools to approach the language arts curriculum, independent assignments, and the text in testing circumstances. We call it "visual literacy." Understanding and developing visual literacy is the missing link for many students. Visual literacy is when students can *make sense of the content* of a lesson and *develop meaning* through nonlinguistic lenses. Evidence of their learning occurs when students are asked to respond. They can communicate their understanding appropriately in speaking, writing, and assessment situations.

In any curriculum, students need to develop listening so they can eventually produce speaking and writing. Listening and reading are *receptive* processes, while speaking and writing are the *productive* functions in language. The Word Cards facilitate *productive* use of language and promote interaction.

Other nonlinguistic tools recommended by the research are graphic organizers, doodles (from both students and teachers), realia such as bringing artifacts and "real-world" applications into the classroom, and modeling of tasks and lessons. These tools help students to *see* what is needed and will be learned. The students report that they like taking "snapshots" of words and concepts. Seeing makes listening and "producing" easier.

How the Word Cards Intervention Works

Here are some of the best ways to use the Word Cards:

METHOD ONE: Say It – Show It – Spell It

One of the simplest ways to use these cards:

1. **Show the Word Card with the printed word covered** with a piece of sentence strip.
2. **Ask the students what they think the word might be**. Play with the students about the word. If it is a unique word, you might use a game

of hangman to create a "muscle memory" for retrieval for the word. Let the students guess the word, predict what the word might be, or come up with synonyms for the word.

3. **Tell the students the word and show the card** in its entirety.
4. **Say, "SAY it."** Wait for a choral response of the word. Ask for it at least three times.
5. **Dialogue about the word**.
6. **Provide the students** with a simple kinesthetic definition of the word.
7. Say, "SHOW it."
8. **Spell the word** for the students. "SPELL it." Define it. Show it. Go through the cycle by calling it out.
9. "Boys and girls, Say it. Show it. Spell it."
10. **Repeatedly call it out** and ask for it in different sequences.

METHOD TWO: Telling Stories and Compare & Contrast

(You must be the judge of whether or not the word cards will lend themselves to this activity.)

1. **Select a series of six to eight cards** and, based on the pictures alone, work with the students on telling stories.
2. **Remind the students of the theme** and how stories are constructed in the unit.
3. **Let them take some time or work with a "buddy"** to outline or develop a story line using the word cards as prompts.
4. **After stories have been created and shared,** provide an outline of the real upcoming anthology to be read in the theme.
5. **Compare and contrast the stories** made up from Word Cards and the actual story. Ask students, "Did the Word Cards reveal any aspect of the story?"

METHOD THREE: Straight Vocabulary

Another technique for the word cards is explicit vocabulary instruction preceding the experience of viewing the word in the story. This kind of frontloading is recommended by Susanna Dutro and is a significant part of the English Language Learners' arsenal for preparing for access to the core language arts program. There are five aspects of vocabulary knowledge that students need to have in order to meet the word in text and not be slowed while reading.

1. **Immediate recognition** of the word and an image with the meaning attached.
2. **Sense of the word in terms of part of speech** and usage.
3. **Ideas about how to check for the context clues** in the text to understand the use of the word in the current narrative.
4. **Student places value on the word and recognizes** that it has an important role in the narrative or text.
5. **Word has been tagged as one that must be retrievable** for assessment or future use.

METHOD FOUR: Icons

One of the students' favorite activities is the exchange of an actual Word Card for an icon.

1. **Have the students look at the Word Card** and then make up an icon for the word.
2. **Then have students write a sentence** using the icon instead of the word, just like hieroglyphics.
3. **Have students trade off to see if they can "read" the icons their peers** have made for the Word Cards in a vocabulary lesson. This really forces the students to use context clues and to understand the word. Besides, it is great fun!
4. **You can make the lesson more difficult** by taking the Word Cards down or turning them around so that they can't be used during the activity.

K and ELL Scaffolding:

Have students cut pictures out of a magazine to use as the basis for new ideas and words that they want to learn.

Why the Word Cards Intervention Works

This intervention works because it provides explicit opportunities to teach new words. The components include: lesson presentation, structured or guided practice, independent practice, and progress assessment. Overall, direct and explicit instruction in Word Cards should follow a schema of teacher-led, teacher-student, and student-student actions that lead to long-term memory retention of the vocabulary words and an understanding of how to use the words in multiple settings (Figure 2.12).

Figure 2.12 **RTI Strategies Scaffolded Practice**

I Do It: Teacher presents content and leads students in application activity.

We Do It: Students work with the teacher in less-structured whole-group practice.

You Do It: Students work with students to check personal understanding.

We Check It: Formal and informal assessments are done for the students and for the teacher to ensure ongoing monitoring of progress, standards acquisition, and proficiency.

Students need to have a clear picture of the outcome of any assessment. It should provide them with specific information about what they need to work on and what they are doing well. Students should know what they already know and what they need to know.

Progress Monitoring for Word Cards

This strategy has a variety of methods for helping students learn new vocabulary and clearly see new concepts they are learning.

- If students struggle to connect words to pictures on the Word Cards, ask them what other pictures would help them remember the word.
- For students who are slow to learn the vocabulary on the Word Cards, work with them in a small group and go through the cards quickly with high repetition.
- Have students draw their own cards or color in their own cards to help them make a connection to the word and the picture they have helped draw or color.

Have fun with the students as they review the Word Cards, make Word Cards, and devise pictures that represent words and the concepts they represent.

Listening Intervention #3: Fist-to-Five

Learning new academic vocabulary is one of the quickest and most effective ways for our students to gain the building blocks of new knowledge. There is a very strong correlation between word knowledge or vocabulary and the overall conceptual knowledge that students develop. Measuring our students' vocabulary knowledge is a fast way to assess their overall conceptual knowledge. Moats (2000) points out:

> Knowledge of word meanings is critical to reading comprehension. Knowledge of words supports comprehension, and . . . networks of words, tied conceptually, are the foundation of productive vocabulary.

Individual word knowledge builds the knowledge of larger concepts or schemas. Look at the basic words that may support the concept of a pet:

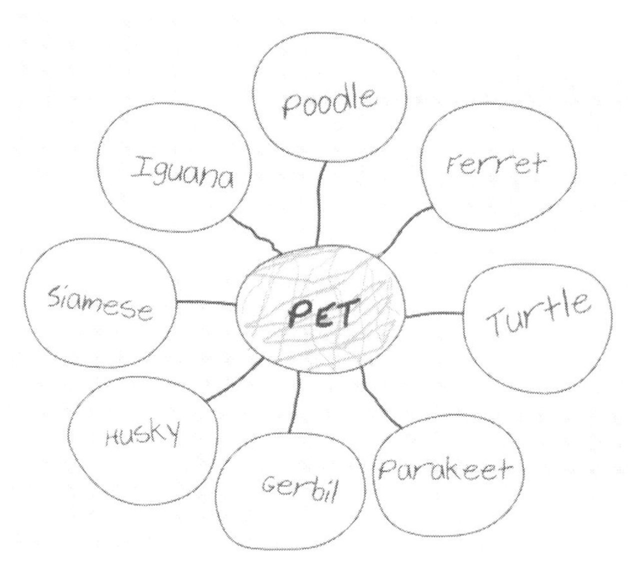

Let's take a different look at vocabulary that connects concepts to meaning for students. Instead of looking at types of pets, let's look at words that describe what our pets may mean to us.

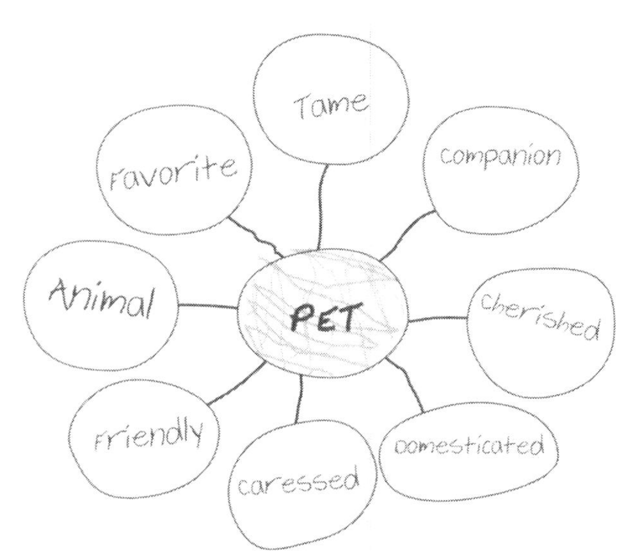

When our students can understand a variety of connected words, the vocabulary takes on meaning and the students can quickly develop conceptual knowledge.

What Fist-to-Five Looks Like

Young students in K-2 are confronted with new words every day. We often state things in terms of our students know a word or they do not know a word. The reality is that our students have different levels of understanding of particular vocabulary words. Helping students see the different levels of understanding is important in encouraging them to gain a complete understanding of words. Caldwell and Leslie (2005, p. 39) talk about the importance of word knowledge and constructing meaning: "What does this mean for intervention? Comprehension involves knowing the meanings of words and using them to construct meaning."

It is important that our students are aware of the level of understanding they have for the words they encounter as they learn. Young students can quickly determine their level of understanding by assessing words according to the following five criteria (Figure 2.13 on page 40).

Students can rate their level of understanding according to these five levels, to help them know where they need to learn more and what they already know. For most students, it takes 12 to 14 reoccurring exposures to develop understanding at the (I know it!) Level 4.

Figure 2.13 **Five levels of Understanding Word Knowledge**

1. I have never seen the word before.
2. I've heard of the word, but I don't know what it means.
3. I recognize the word—I think it has something to do with...
4. I know the word in one specific context.
5. I know the word in several different contexts!

The Fist-to-Five strategy is a strategy that can be used daily to quickly assess your student's level of understanding of key vocabulary words. A list of 10–12 words can be presented to young students and within 5 minutes you will receive feedback regarding those words students need intervention support.

How the Fist-to-Five Intervention Works

This strategy is a quick and seemingly simple intervention that provides students and the teacher with quick and comprehensive feedback regarding vocabulary knowledge.

1. Write a **list of academic vocabulary on the board** that they know will be part of the day's lesson (i.e., words like addition).
2. **Pronounce each word** out loud, making sure to enunciate each word.
3. Ask students to **say each word out loud** together in chorale fashion.
4. Ask students to put their **fist up in front of their chin** in the "ready position."
5. Say a word and have **students assess their understanding** of the word.
6. Students then should show their level of understanding by **holding up one, two, three, four, or five fingers** to indicate their level of knowledge.
7. Quickly **scan the room** to note whether students are holding up a majority of low numbers (one and two fingers) or a majority of high numbers (four and five fingers).
8. Ask a student who is holding up a 4 to 5 to **provide their definition** of the word.
9. **Clarify the definition** as needed and then move onto the next word.

> **K and ELL Scaffolding:**
>
> In addition to spelling the word on the board and phonetically sounding out the word for students, show or draw a picture of what the word may look like for students to help activate their knowledge of the word.

Why the Fist-to-Five Intervention Works

This strategy works because students can be introduced to a variety of new words in a very short amount of time. They can evaluate their understanding of the words and signal their understanding to the teacher. The teacher gets a very quick idea of what words the students know, and which words may need more attention to develop comprehension. Teachers can then provide background information and examples to build on student understanding. This intervention works because asking students to raise their hand and hold up the number of fingers that matches their level of understanding allows us to assess understanding in mere seconds, and then to adjust our instruction accordingly.

Progress Monitoring for Fist-to-Five

This strategy is extremely strong for progress monitoring, because it allows teachers to quickly monitor students' progress in their vocabulary development.

- Pay special attention to students that are consistently holding up 1s and 2s, and give them additional support.
- Use the quick feedback from Fist-to-Five to adjust the proper amount of time devoted to learning new words and definitions that will be needed for effective comprehension of the lesson.
- Check several students' understanding to make sure that students are holding up a finger that accurately identifies their understanding

This strategy is a critical strategy for monitoring the progress of our students' level of understanding on key vocabulary terms.

Listening Intervention #4: Academic Language Interventions

The achievement gap affects many of our students, particularly students from poverty and second language learners. As noted in the introduction to this chapter, a language gap exists for many students before they even enter kindergarten. Taylor (2006, p. 28) emphasizes the importance of instructing students in academic language.

> Teachers must be careful to directly instruct academic language, use academic language in class, and expect students to use academic language so that they own and understand the language when it is encountered during assessment or independent content reading.

Academic language is the underlying language of school that students need in order to be successful in school. It provides the essential components of constructing knowledge and learning in all of the content areas.

What the Academic Language Intervention Looks Like

Language is the medium or method in which we convey conceptual knowledge. As we understand how academic language is organized, it will help us better understand how knowledge is organized. Academic language serves as the glue that helps students transition. Without a basic knowledge of grade-level academic language terms, then students may become lost and overwhelmed by the rigor of school.

The analogy of the builder—students find it helpful if we compare constructing knowledge to the building process. Students are the key factor in building their knowledge, and they need bricks and mortar to build a proper house of learning. Academic language has two parts that are essential to learning at school—the bricks (content knowledge language) and the mortar (general academic language). All students need to learn both the bricks and mortar of academic language so that they can effectively build their own knowledge structures and become successful learners throughout their lives.

Bricks – Content Language

The first kind of academic language we will discuss is the content knowledge language or bricks of learning. Mathematics, language arts, science, and social studies serve as the four cornerstones of an academic learning. The specific content language in the core subject areas also act as bricks in building knowledge structures for learning. Most teachers spend a fair amount of time explicitly instructing students in the key language terms that are building blocks for content learning. Here are a few examples of standard K-2 content knowledge terms or the bricks of learning.

- **Mathematics** (i.e., *addition, equals, subtraction, number, and sum*)
- **Language Arts** (i.e., *capital letter, sentence, period, spelling, and noun*)
- **Science** (i.e., *observe, samples, experiment, identify, and collection*)
- **Social Studies** (i.e., *community, history, government, immigrate, and election*)

Mortar – Academic Language

General academic language is the mortar that helps students to build structures of knowledge. These words are an essential foundation and they bind the bricks of content knowledge language together. Academic language is the mortar of a knowledge foundation and it holds a knowledge framework. Mortar serves as both the foundation for learning as well as the binding agent that holds the bricks of learning together. When K-2 students learn the key words of academic language, then they have a foundation to build their core structures of knowledge in math, science, language arts, and social studies. There are three types of general academic language that act as the glue that cohesively connects learning together.

- **Action Words** (i.e., *analyze, define, increase, compare, and label*)
- **Transition Words** (i.e., *if, again, during, prior, and because*)
- **Concept Words** (i.e., *role, goal, emotion, target, and energy*)

Academic language is the mortar of a knowledge foundation and it holds a knowledge framework. Mortar serves as both the foundation for learning as well as the binding agent that holds the bricks of learning together. Learning the essential academic language at each grade level will help students connect their learning successfully and avoid becoming an academic casualty. Fully developing our students' academic language will dramatically boost the coherency of the K-2 curriculum while integrating the core subjects of school. Johnson (2009, p. 24) observes:

> Academic language is the powerful language of learning and making meaning. In many ways, academic language provides the common thread that is found in all core content areas and woven throughout core standards, curriculum, and assessment.

Let's take another look at a chart from the California Department of Education (2007). Notice how Academic Language is a critical component in connecting a students' ability to decode or "learn to read" and their ability to comprehend or "read to learn" (Figure 2.14).

Figure 2.14 Framework for Reading

DECODING Learning to Read					COMPREHENSION Reading to Learn				
Word Recognition Strategies				Fluency	Academic Language		Comprehension Strategies		
Concepts of Print	Phonemic Awareness	Phonics	Sight Words	Automaticity	Background Knowledge	Brick and Mortar Vocabulary	Syntax & Text Structure	Comprehension Monitoring	(Re)organizing text

Adapted from John Shelfbine/Developmental Studies Center

Students who have a foundation in academic language are able to make the transition from learning to read (decoding) and reading to learn (comprehension) and avoid getting behind in school.

How Academic Language Intervention Works

We need to provide explicit instruction in both the content language and the academic language of school.

1. **Explain to students the difference between content language-bricks and the academic language-mortar.** (Even very young students can understand this analogy and will benefit from its insights.)
2. **Provide example of content language-bricks and academic language-mortar words.** (A full list of academic language-mortar words for each grade level is available in *Academic Language! Academic Literacy!* by Johnson, 2009.)
3. **Before lessons, identify the key content language-bricks that will be used in the lesson** (i.e., a lesson about weather may include brick words like "sunny," "snow," or "clouds").
4. **Integrate the academic language-mortar words into each of the four subject areas** (i.e., the mortar word *analyze* could be introduced into analyzing the rhyming patterns in language arts, analyzing the weather in science, analyzing the different groups in social studies, or analyzing patterns of numbers in math).
5. **Use the Fist-to-Five intervention strategy discussed previously to see how many brick and mortar words students know.** (This is a fast and fabulous way to assess students' word knowledge.)
6. **Use the Academic Vocabulary Graphic Organizer intervention strategy discussed next to more explicitly instruct students in the brick and mortar words.** (You will find this strategy to be very thorough in developing students' word knowledge.)

K and ELL Scaffolding:

Provide students extra small-group instruction to review and learn in depth the key academic language terms that will hold their learning together.

Why Academic Language Intervention Works

Students who lack a foundation in academic language seldom have the capacity to grasp fully the grade level concepts essential in the "learning to read" phase. Without a firm foundation of academic language to build upon, students' knowledge structures can become unstable with cracks in their learning. If students lack the academic language to serve as the mortar to hold the specific content bricks together, then gaps will begin to show up as students' progress into the upper elementary grades. Without the mortar to hold the bricks together, then everything can fall apart and learning chasms will eventually appear. Students who have cracks and gaps that are never filled with the mortar of academic language are prime candidates for eventually falling into the chasm of school dropouts. Students who develop academic language effectively go onto succeed in high school and they have the language resources to succeed in college and career.

Progress Monitoring for Academic Language

Because we take academic language for granted, it is important to really hone in on our students' understanding and ability to use the brick and mortar words that build knowledge structures.

- Take time to make sure that students learn new brick words in different subject areas and pull students into small groups if they need extra help.
- Train instructional aides in the essentials of academic language so they can pay special attention to students' knowledge of mortar words that help glue the bricks of learning together.
- Give students plenty of opportunities to speak and write using the mortar words of Actions, Transitions, and Concepts that glue or connect words, phrases, and sentences into effective communication.

As you assess academic language frequently, your students will start to recognize how important these words are to their own learning. If we provide our students with the essential bricks and mortar of learning, then our students will have the necessary resources to build a skyscraper of success that can reach as high as they choose to go.

Listening Intervention #5: Academic Vocabulary Graphic Organizer

Explicit instruction in academic language works because it provides students with the essential resources and building blocks for independently constructing knowledge. Students with low socioeconomic backgrounds and limited language will struggle throughout their school careers if they never develop the academic language needed to effectively construct grade-level knowledge. Students need to learn the language or vocabulary of school. Evans and Sorg (2007, p. 4) highlight the importance of academic vocabulary and learning:

> Vocabulary knowledge is one of the most reliable predictors of academic success." The authors go on to say, "Knowing academic vocabulary—the "vocabulary of learning"—is essential for students to understand concepts presented at school.

Unless many students are provided intervention strategies that work, they will struggle with the vocabulary gap that by grade three becomes a reading gap as "reading to learn" becomes so important. Archer's (2003) Vocabulary Development (http://www.fcoe.net/ela/pdf/Anita%20Archer031.pdf) cites, "In many ways the 'Reading Gap,' especially after second and third grade, is essentially a Vocabulary Gap—and the longer students are in school the wider the gap becomes."

Students need systematic instruction in academic language and the vocabulary that helps them construct knowledge. A comprehensive method for explicitly instructing students in academic vocabulary can help all students and particularly struggling students with a consistent format for learning the essential language of school.

What Academic Vocabulary Graphic Organizers Look Like

The academic vocabulary graphic organizer provides both the teacher and students with a powerful tool for analyzing and learning academic vocabulary. The graphic organizer should be used several times a week, and more often in small group interventions for struggling students. Remember students need 12 to 14 reoccurring exposures to a word in order to develop their understanding at a Level Four (I know it!). K-2 students should include all five of the following components as they develop their word knowledge of important academic vocabulary.

1. **Write the Academic Word** (Students should say the word out loud auditorily. Write it down kinesthetically, and then view it visually.)
2. **List Synonyms** (Writing down similar words will help describe and connect the word to the student's prior knowledge.)
3. **Draw a Picture** (The act of drawing will provide a symbol for remembering.)
4. **List Examples** (Writing down examples will provide a real-world context for the word.)
5. **Write down a working definition** (Students who generate their own definition remember the word much better.)

Figure 2.15 is an example of an Academic Vocabulary Graphic Organizer designed by first-grade students. Notice the working definition, synonym, picture, and example.

As students at the earliest stages (kindergarten and even preschool) learn to write down the word, think of a synonym, draw a picture, list examples, and come up with a working definition, they will begin to understand how important vocabulary and language is to their academic success.

How Academic Vocabulary Graphic Organizers Work

It works by asking students to raise their hand and hold up the number of fingers that matches their level of understanding. Begin by asking students to do a "Fist-to-Five" and share their level of understanding.

1. ...if they have never heard the word before.
2. ...if they have heard the word, but don't know what it means.
3. ...if they recognize the word and think they know what it has to do with.
4. ...if they know the word in a specific context.
5. ...if they know the word in several different contexts.

Academic vocabulary words that have a high number of 1s and 2s that are also key brick or mortar words should be explored in detail by working with a graphic organizer.

1. Give students an **academic language term** to learn in depth.
2. Put them in **partners** so they can help each other.
3. Ask them to **write the word in the center** of the graphic organizer.

Figure 2.15 **Academic Vocabulary Graphic Organizer Sample**

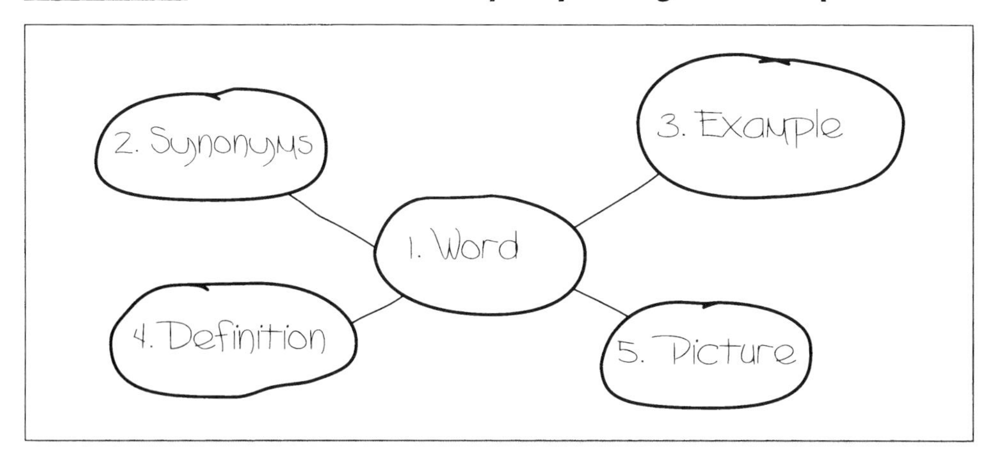

4. Ask students to work together to come up with **synonyms, a working definition, and examples**.
5. Ask students to **draw a picture** or symbol that represents the word.
6. Have students **share** their answers with the class.

> **K and ELL Scaffolding:**
>
> The teacher will need to go through the academic vocabulary graphic organizer with students as a group. Students can work collectively together to come up with answers to the graphic organizer. The teacher can help students to think of synonyms, examples, and a working definition.

Why Academic Vocabulary Graphic Organizers Work

This strategy works because students can see what they hear! It helps children organize their thoughts and what they know. The more graphic representations children learn, the better they are at focusing their listening to discern what they will need to know to fill out the graphic map. When students work in pairs, they negotiate their working definitions, examples, and synonyms of the words. The process of negotiating helps students retain information better and they extend their learning to higher levels. Students will also expand and elaborate their learning of these important academic vocabulary words by listening to the answers of their peers. After working with this graphic organizer for several months, students will start to automatically try to figure out examples, working definitions, synonyms, and pictures of new words they encounter in context. In this way students will become effective learners of new academic vocabulary.

Progress Monitoring for Academic Vocabulary Graphic Organizer

This strategy is extremely strong because it allows teachers to quickly monitor students' progress in their vocabulary development.

- Look at the working definitions that students come up with.
- Recognize that students who have difficulty drawing a picture of a word may also have difficulty visualizing new words or concepts in full detail.
- Keep track of how many new words young children are explicitly adding to their vocabulary every week.

As students expand the number of words they actively add to their working vocabulary and they have a consistent process for understanding new words, then they will become better independent learners of new words.

Summing It Up

Although listening is the literacy activity that students spend the most time engaged in at school (at least 50% of the day), it is rarely ever directly targeted for improvement. Students need intervention listening strategies to help them learn how to provide their attention to academic purposes. All students will benefit from strategies that support active listening, while many students will definitely benefit from additional small-group interventions devoted to improving listening strategies. When students are able to identify the purposes of learning or listening activities, they will be able to direct their attention more effectively. Using pictures to engage word development in an inductive fashion will benefit students. The Fist-to-Five intervention is a favorite of students, because they get to actively respond and it provides immediate feedback to the teacher to direct future instruction. Academic language provides both the essential bricks (content knowledge language-language arts, math, science, and social studies) and the mortar (general academic language-actions, transitions, and concepts) for our students to become builders and construct their own internal knowledge at school. The Academic Language Graphic Organizer can increase our student's vocabulary so they can make more meaning from classroom conversations and content. Increasing listening strategies will increase our student's attention, motivation, and learning.

Reflection

1. Do your struggling students have ample opportunities to acquire new language that will make them effective builders of their own learning?
2. Which listening strategies do you believe will help to develop your students' listening and learning abilities most effectively?
3. How does learning vocabulary help our students be more effective listeners and creators of their own meaning?

3

K-2 Intervention Reading Strategies

"The more that you read,
The more things you will know.
The more that you learn,
The more places you'll go."

—Dr. Seuss

Jeremy walked down the main hallway of his new school. He looked around at his third school in three years. Going into the second semester of second grade felt like a tough time to be attending a new school. He had just started to feel comfortable with his previous teacher, Ms. Watts. He rarely read books, even though he liked to look at the pictures. It always seemed that he never quite had enough words to express his thoughts, so he just tried to blend in. It always seemed he just slipped through the cracks. He played by himself at recess time, and usually ate lunch alone in the cafeteria. Now that he was at a new school, the pattern felt like it was going to start all over. It seemed that Jeremy attended a new school so often that it was difficult for him to transition to learning everything that he so desperately needed to learn. He wanted to fit in at school, yet it always felt like he was behind in class and at the end of the line in everything that he did. He lacked confidence academically and socially. Jeremy is a student in transition, and he is a struggling reader.

The Language/Literacy Connection

Our understanding of language is at the heart of our ability to be literate individuals who can effectively learn. The language that our students know and can produce provides a foundation for learning, while the literacy that our students can produce provides a framework for their learning. Students who have more words in their vocabularies have more comprehension when they read. Vice versa, students who read more add more words to their language vocabularies. Reading feeds vocabulary development, and in turn vocabulary development feeds reading comprehension. Frequent readers add more words to their vocabularies, and the larger vocabularies mean they can read more fluently. Vocabulary has a very direct connection to reading level. A student's ability to comprehend what they read is impacted by the amount of vocabulary they have and their ability to seamlessly, rapidly, use their vocabulary. Vocabulary has a very direct connection to level of knowledge. Knowledge structures, cognitive understanding, schemas, or whatever one wants to call them are built with words. Vocabulary is the quick and easy way to assess knowledge. The more vocabulary someone has about a certain topic or area of knowledge, the more it seems they know about the topic. For example, it is said that Eskimos have at least six different words for snow, while most people only have one. Eskimos are experts in their knowledge regarding the topic of snow, and so they have a larger more robust vocabulary in this area. Individuals that know a variety of specific vocabulary about a topic typically have more detailed, specific knowledge of the topic. There is a direct connection between academic language (vocabulary words related to school) and the ability to negotiate learning through academic literacy (reading, writing, speaking, and listening processes) (Johnson, 2009). Students who have good academic literacy also have good learning skills (retrieval, organizing, attention, synthesizing, etc.) Simply stated, students who develop academic language are able to develop their academic literacy skills and they become better academic learners.

The Language/Literacy/Learning Connection

For example, basic IQ (intelligence quotient) tests often measure intelligence by assessing a series of vocabulary words. Also, teachers who have taken the GRE (graduate record exam) to get into graduate school may have noticed that a significant portion of the test (about two-thirds) is based on academic language or vocabulary knowledge. So, many formal and informal assessments use vocabulary to measure knowledge because of the strong correlation between knowledge and vocabulary. Again, we construct knowledge and meaning from language, and the more language we recognize and can comprehend, the more knowledge we typically have.

Recognizing Words (Language) + Understanding Passages (Literacy)

=

Constructing Meaning (Learning)

Honig, Diamond, and Gutlohn (2000) states the relationship between language, literacy, and learning in this way:

> Readers construct meaning from two major sources: words and passages (strings of words). Proficient readers recognize and obtain meaning from words rapidly, effortlessly, and unconsciously—automaticity (p. 13).

Young students in K-2 classrooms need support in the following language and literacy areas (Figure 3.1).

In this chapter we will focus more strategies that develop our students' understanding of the passages they read. Most early core textbooks do a good job of emphasizing the basics of phonics, yet they often do little to develop young students' comprehension. The strategies in this section support and strengthen young students' reading comprehension. So, let's jump into our first K-2 intervention strategy for reading.

Figure 3.1 Keys to Language & Literacy

Recognizing Words (Language)	Understanding Passages (Literacy)
♦ Print Concepts	♦ Comprehension Strategies
♦ Alphabet Recognition	♦ Text Structure
♦ Phonemic Awareness	♦ Independent, Wide Reading
♦ Decoding	♦ Book Discussions
♦ Spelling	
♦ Vocabulary Development	

Reading Intervention #1: Cloze Intervention

The Cloze intervention has helped assess and identify students' reading levels. It has been used as a diagnostic tool to determine the level of understanding for learners. The process of the Cloze intervention requires students to consider context, consider how vocabulary words are used, analyze sentence structure, and a myriad of other very important skills that develop good readers. The Cloze process draws on the learner's knowledge of the content, vocabulary, grammar, and spelling.

What the Cloze Intervention Looks Like

The Cloze intervention is a strategy where words or parts of a word are omitted from a passage of printed text. Strickland, Ganske, and Monroe (2002, p. 106) point out:

> Cloze passages encourage students to use context to figure out unknown words. The procedure is easy to carry out. A short passage of text is selected and copied or summarized on the chalkboard or a transparency. Several words are deleted, and students are guided to figure out the missing words by using the sense of the surrounding sentences.

Students should read and study the passage to determine which words or letters are missing. Here is a simple example of a cloze passage with omitted words.

> A quick and easy way to ___________ out a missing word is to consider the context surrounding the ___________.

Students should do their best to insert the appropriate word into the missing space, analyzing the text to figure out what word(s) will complete the meaning from the text. This intervention is often used as a diagnostic reading assessment technique. Cloze interventions improve students' comprehension and understanding. The purpose of the Cloze intervention is for students to determine the precise word that the author of the text used in the original passage.

Through the process of engaging in the Cloze intervention, teachers will be able:

Figure 3.2 Five Ways the Cloze Intervention Can Be Implemented

1. You can **omit only the first letter** of the word.

   Johnny ate the entire piece of birthday _ake.

2. You can **omit every letter except for the first letter** of the word.

   The ball bounced off the backboard and went into the b_ _ _ _ _ _.

3. You can **omit the word and provide choices** for students to select.

   The lion has ________, and the bird has ____________.
   (fur, fins) (hair, feathers)

4. You can **omit the word and provide the proper number of letter spaces.**

   Manny grabbed the bow and shot the _ _ _ _ _.

5. You can **omit the word and just leave a blank space** without providing the number of letter spaces.

   The yellow school bus pulled up to the curb and the _________ got onto the bus.

- To identify students' knowledge of a subject and understanding of the reading process
- To assess the extent of the students' vocabularies
- To encourage students to think critically and analytically about text and the content
- To encourage students to make meaning out of written text

This intervention works extremely well with interesting and engaging text. Of course any text can be used, yet the intervention works extremely well with informative text. The intervention encourages risk taking as students venture out with a possible solution (Figure 3.2).

Clozes help students learn to look for context clues. Although context clues are typically outlined in most core K-2 textbooks, it is worth noting four key types of context clues (Figure 3.3 on page 58).

K and ELL Scaffolding:

The Cloze provides the support that students need as they collectively fill in missing letters or words. For beginning readers, start by omitting the first letter to help students see how letters added to other words can create words.

Figure 3.3 Types of Context Clues

1. **Restatement or Definition** – Another way of saying or defining the word.
   i.e., The professor, who was my teacher in college, wrote me a letter of recommendation.
2. **Similar or Synonym** – A word or phrase with a meaning that is the same as, or very similar to, another word or phrase.
   i.e., The cranium or bone structure that protects our brain is very strong.
3. **Opposite or Antonym** – A word which has the opposite meaning to another.
   i.e., The monkey's calloused foot felt rough in contrast with its soft hair.
4. **Explanation or Example** – A statement that describes a set of facts or which attempts to show how or why something is the case.
   i.e., The dolphin is a mammal because it breathes and provides its young milk.
5. **Analogy or Logic** – A logical statement following the word *like* describes the word.
   i.e., An exoskeleton is like a suit of armor that protects the ant.

How the Cloze Intervention Works

To prepare materials for Cloze exercises, any of the following techniques may be used:

1. **Select a passage of text** that is the appropriate length for the students' grade level. The passage should be easy for students to read and understand.
2. **Make sure to keep the first and last word** in the sentences and all punctuation intact.
3. **Select the words that you will omit from the passage using a word-count formula,** such as every seventh word or other criteria.
4. When preparing the final draft of the passage, **make all blanks of equal length** to avoid including visual clues about the lengths of omitted words.
5. **Students should read the entire passage** to get as much context and meaning as possible before they begin to fill in the blanks.
6. **Students should do their best to fill in each blank** with the word they feel is best.

7. Although there should be **no time limit** for this exercise, please record the time that it takes for students to finish.
8. After students have filled in the blanks, **they should reread the passage** to make sure they like their answers.

Cloze Assignment

Read the passage below with the missing blanks. Please fill in the blanks with the letter or words that the author intended to be in the blank space. Every correct answer is worth one point.

Example: With the price of f______ going up all the t______, more people are trying t___ raise some of their f_____ in their own back y______.

When the learners do well with this task, indicate only the blank with no additional clues. Accept any word that seems a reasonable fit:

Example: Instead of grass, you can _______ rows of lettuce, tomatoes, _________ beans.

Answers: (food, time, food yard) (grow, and)

To assess students' knowledge of the topic or their abilities to use semantic cues, delete content words that carry meaning, such as nouns, main verbs, adjectives, and adverbs. To assess students' use of syntactic cues, delete some conjunctions, prepositions, and auxiliary words.

Why the Cloze Intervention Works

The Cloze intervention helps students to think critically about text and analyze the organization and choices made by the author. This strategy provides teachers and students a chance to discuss language and vocabulary. The Cloze strategy helps students make a prediction, infer meaning, consider context clues, analyze word choice, and a variety of other processes that develop comprehensive understanding. The strategy helps students focus in on words, rather than just skip by words without processing meaning. Many students decode text, without comprehending individual words or making meaning from the entire passage of text.

The Cloze strategy can be used to:

- Determine readability of text
- Strengthen cuing systems (syntactic, semantic, and graphophonic)
- Help determine background knowledge of a topic
- Help with comprehension

This intervention may be effective because it makes explicit to the reader so many of the processes that he or she usually processes without having to stop and think. This intervention works because the activity can be scaffolded to a difficulty level that matches the students' reading level. Just a letter or two can be omitted or many words can be omitted to check students' ability to comprehend and make meaning. This intervention can be matched up with other interventions strategies like summarizing and so on. This intervention engages students in all of the internal cognitive processes that good readers use when reading. A measurable score on each passage can help track student progress over time.

Cloze Progress Monitoring

Typically a passage needs to have at least 20 words missing to get an accurate score (Figure 3.4).

- Provide students with grade level Cloze passages with 20 missing words at least every six weeks.
- For students who correctly fill in less than 40% of the missing words, provide extra help identifying context clues.
- Give extra vocabulary support to students to develop their word knowledge.

Figure 3.4 **Cloze Scoring Levels**

60% or higher correct	**Independent Reading Level** (students should be able to read on their own)
40% to 60% of words correct	**Instructional Reading Level** (students can read with teacher's help)
Less than 40% of words correct	**Frustration Reading Level** (too hard for students, even with help)

Reading Intervention #2: Inferring

Inferring is a matter of bridging the gap. At the earliest levels of learning to read the students will replace a letter to change the word to a different meaning. Eventually students can begin to infer from short sentences. Eventually we want students to infer from paragraphs and between paragraphs. It is difficult to include every bit of information in the text. We write in ways that cause learners to create their own meaning from the context of the message. Actual cognitive strategies can be taught to students so that they become better at inferring from reading. The ultimate objective of the instruction process is to develop comprehension regarding the world around us. Comprehension is developed as we understand the logic structures that organize our language, and we also understand the logic structures that organize our conceptual knowledge. Duffy (2009, p. 122) notes:

> Virtually all comprehension strategies involve inferring in the sense that comprehension requires readers to note text clues, to access prior knowledge associated with the clues, and then, on the basis of that background knowledge, predict (or infer) what the meaning is. So, in this sense, inferring is something a reader does as part of all comprehension strategies.

Inferring is the process by which we recognize the meaning of words. It is important for learners to infer from both oral and written statements. The challenges become much bigger for learners when they engage with written text, because formal written text is more academically complex and dense. The differences learners face, when inferring from oral statements and written statements, stems from the fact that writing has much more challenging academic language and much more complex language structures compared to spoken communication. Readers who effectively increase their inferential skills are able to draw accurate conclusions and make sense of the material they read.

What the Inferring Intervention Looks Like

As they advance in grade levels, students face reading assignments and textbooks that are cognitively more difficult. The cognitive challenges of textbooks seem to perpetually increase because of an increase in the academic demands of language, an increase in the logical complexity of concepts, and an increase in the structural complexity of language.

> Students learn approximately 90% of new words outside of school

All of these challenges heighten the importance of increasing students' abilities to infer meaning from words and make logical connections between ideas. Students who know academic language contained in assigned reading passages are able to understand reading assignments more quickly and to learn key concepts more effectively. The research shows that approximately 90% of words learned by students are learned outside of explicit instruction. The following numbers adapted from Hirsch (2006) and Nagy and Anderson

Figure 3.5 Words Learned by Students

Reader's background	Average number of words learned per day	Approximate number of words learned per week	Estimated number of words learned per year
Socioeconomically disadvantaged students	7	50	3,000
Working class students	12	85	5,000
Professional students	14	100	5,500

(1984) show the comparison in the number of words learned by students from different socioeconomic backgrounds (Figure 3.5).

It is important that students learn the various types of general inferences that are made as they consider new information. The following six types of inferences should be developed and understood by all young students (Figure 3.6).

Figure 3.6 Six Types of Inferences with Examples

- **Infer Characteristics of an Object** – *The second-grade student shot the ball through the hoop.* (We might infer the ball is an orange basketball.)
- **Infer Time** – *Julie and her friends waited for the school bus to pick them up while the sun rose over the hill.* (We might infer that it is in the morning.)
- **Infer Action**– *Manny hit the ball and ran to every base before crossing home base.* (We might infer that Manny hit a home run.)
- **Infer Location** – *After the kindergarten students cleaned their table, they emptied the milk cartons on their trays into the trash can.* (We might infer that Julie is in the school cafeteria.)
- **Infer Feelings/Attitudes** – *Bobby looked into the principal's treasure box and selected a new Dr. Seuss book as a reward for perfect attendance.* (We might infer Bobby is feeling happy.
- **Infer Causal Relationships** – *Mrs. Jackson, the nurse, had Tammy's mom pick her up early from school.* (We might infer that Tammy is sick.)

Figure 3.7 Steps to Making Inferences

1. **Read a sentence to students** from an interesting book or have them read sentences.
2. **Ask students to identify an inference** that they could make about the sentence.
3. **Have students classify or categorize the type of inference** they have made.
4. **Ask a student to volunteer** their answer.
5. **Ask other students to share** a different inference or classification for their inference.
6. **Proceed and read other sentences,** stopping at one that provides a good inference.

When students have even a few breakdowns in their ability to infer the meaning of specific words and the logical ideas that connect it to the phrases and sentences surrounding these words, then their ability to make sense of the paragraph or passage is limited and their comprehension is compromised. Students who can accurately decode language by recognizing sounds and fluently produce the sounds used for "learning to read" do so by reading simple easy to comprehend stories. As students progress into the upper grades of elementary school and into middle school, the importance of "reading to learn" becomes an increasingly more difficult ability to master.

How the Inferring Intervention Works

This strategy works because it helps students learn the cognitive processes that improve their ability to make connections (Figure 3.7).

The following sentences can be used as a model exercise that will help students recognize the many inferences that they can make as they read (Figure 3.8 on page 64).

K and ELL Scaffolding:

Read sentences to students that have quite explicit inferences that are directly stated in the sentences.

Figure 3.8 **Sample Assignment for Developing Inferences**

Read each sentence and make an inference about the possible characteristics, actions, relationships, and so on and decide which types of inference each sentence exemplifies. After completing, please discuss your answers with a peer.

1. Jeffrey fell off his bike.
2. Someone shook the can of soda pop.
3. David received a smiley face on his paper.
4. Pam received a certificate from the teacher.
5. The milk carton and a puddle of milk were on the floor.
6. Mrs. Jackson reminded all of the kindergarten students to get their coats, boots, and umbrellas before they went outside for recess.

Why the Inferring Intervention Works

This strategy is critical for students to begin reading between the lines and understand important information that the author does not specifically provide. Students infer naturally, yet typically they are unaware of the inferences they are making. Explicitly identifying inferences helps students become more proficient at recognizing inferences. As inferences are shared with one another, students will hear the inferences that their classmates share and they will expand their understanding and broaden their ability to make inferences in the future. Because inferring is the process by which 90% of words are learned by students, practicing this skill can dramatically increase the number of words learned by students in the future.

Progress Monitoring for Inferring

Students can quickly make these inferences as a group and share them collectively with the teacher.

- Check to see that students are using all six different types of inferences as they categorize sentences.
- Check to see if students can make more than one inference for a particular sentence.
- Ask students to share the thinking behind the inferences they make.

Reading Intervention #3: Cognitive Reading I

Learning new information by reading from books can be a cognitively complex process. Many K-2 students, when they read written words, decode them without much comprehension. Cognitive reading strategies are researched methods that increase young readers' abilities to strengthen their comprehension. Rosenshine (1997, p. 201) outlines the benefits strategies play in strengthening the processes of reading: "A cognitive strategy is a heuristic or guide that serves to support or facilitate the learner as she or he develops internal procedures that enable them to perform the higher level operations."

Good readers actively engage their internal cognitive processes when they read; on the other hand, struggling readers are often unaware of methods for engaging their cognitive abilities.

What the Cognitive Reading I Intervention Looks Like

The cognitive reading strategy is actually a conglomeration of several strategies that work together to engage students' reading abilities. These strategies provide a framework for our young students' learning. Poor readers often develop only partial or fractured comprehension because they have difficulty making connections, visualizing, and predicting that kick-start thinking and learning. There are six power-packed cognitive reading strategies that get young students engaged in their reading (Figure 3.9).

In this section we will consider the first three and in the next strategy section we will consider the last three. Look over the list of cognitive reading strategies that initiate young readers' comprehension (Figure 3.10 on page 66).

Figure 3.9 **Cognitive Reading Strategies**

- ♦ Accessing Prior Knowledge
- ♦ Making Predictions
- ♦ Visualizing
- ♦ Determining Importance
- ♦ Generating Questions
- ♦ Monitoring and Clarifying

Figure 3.10 **Beginning Cognitive Reading I Strategies**

- Accessing Prior Knowledge – Make connections to information already known
- Making Predictions – Declare in advance or foretell outcomes
- Visualizing – Form mental pictures of characters, events, or scenes

Maanum (2009, p. 177) notes, "Predicting, visualizing, and connecting are before-reading strategies used in activities to spark an interest in reading. These strategies help to activate prior knowledge..."

Cognitive reading strategies support students as they engage in the processes of reading and learning. Students need to pack their reading activities with suitcase of strategies that will engage their internal cognitive thinking processes.

How the Cognitive Reading I Intervention Works

This intervention works best when students combine the strategies together to strengthen their reading awareness and understanding. Young students may need support in breaking down the vocabulary and identifying the concepts that go into cognitive reading strategies. The conversation around predicting, visualizing, monitoring, and questioning can be rich and helpful for students as they become more aware of the ways good readers strengthen their understanding. We will provide a variety of questions or ideas that will prompt students to engage their cognitive abilities that significantly support reading.

Accessing Prior Knowledge

Learning begins when we connect new information to prior knowledge. The better our young students can connect their prior knowledge to new ideas, the better they will learn. Consider asking the following questions when engaging students in the how of cognitive reading interventions.

- What do you already know about this topic?
- What comes to your mind when you see the pictures in this story?
- What are you reminded of as you read the introduction?
- What can you connect to this new information?

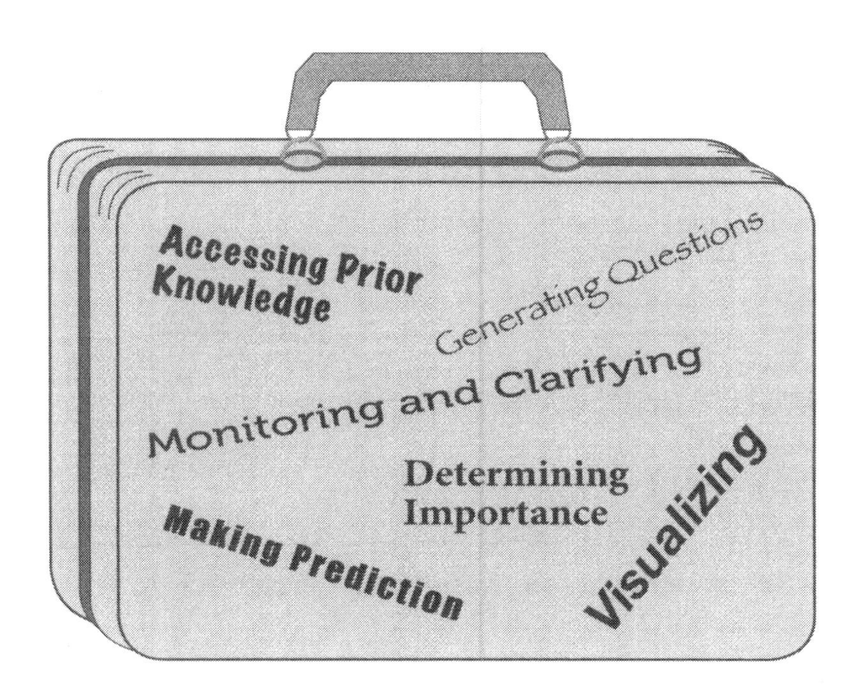

Making Predictions

Making predictions and checking the predictions is a fundamental learning process for developing comprehension. It is a way that students can try to use their prior knowledge and imaginations to make a stab at what will be learned. Predicting is an initial step in making inferences. Good readers constantly make predictions and monitor, revise, and check their learning to see if they understand accurately.

- After looking at the cover of the book, what do you think the story will be about?
- What do you think will happen next?
- If you were to guess the final result from what we have read so far, what do you think will happen?
- How close were your predictions to the result?

Visualizing

Good readers translate the words they read into powerful visual images. Some young students have very vivid imaginations and they can see stories very clearly. Other young students need help engaging their visual abilities as they read. Ask students to consider the following visualization ideas:

- What pictures or scenes come to mind when you read?
- What do you sense, see, or feel when you read?
- What images develop when you read?
- If you made a movie of this information, what would it look like?

> **K and ELL Scaffolding:**
>
> Because young students are just learning to read, reading aloud to students will help you engage their cognitive strategies.

Cognitive Reading I Progress Monitoring

Because cognitive reading processes happen within students' minds, it can be difficult to readily see if students are engaging in these strategies. Asking questions and inviting students to verbally share their thinking with others can give us an opportunity to assess the level of our students' cognitive reading abilities.

- Ask students to share their predictions, visual images, or prior knowledge about a subject as they read.
- Invite your students to do a quick pair-share with their neighbor about what they are thinking about as they listen to you read aloud, and listen to the insights they share.
- Ask students to write down the prior knowledge, visual images, and monitoring processes and keep a brief journal of their thinking processes as they read, and then use these journals to assess their reading abilities.

Reading Intervention #4: Cognitive Reading II

Learning new information by reading from books is a cognitively complex process. Many K-2 students, when they read written words, decode them without much comprehension because so much energy is spent on processing the words. Young students need to take the time to develop strategies that will also develop their understanding or comprehension. Heibert (2009, p. 35) declares, "Engaged reading requires students who are actively using cognitive processes while reading, with an emphasis on the use of cognitive strategies or the development of conceptual knowledge, or both."

As students consciously develop their cognitive reading abilities, they will become strategic readers who get the most out of their reading.

What the Cognitive Reading II Intervention Looks Like

Now that we have considered the first three cognitive reading strategies that help students begin reading, let's look more closely at three cognitive reading strategies that help students while they are reading. Determining importance, generating questions, and monitoring and clarifying are cognitive strategies that help students while they are engaged in the reading process. Let's now look into the Cognitive Reading Strategies (Figure 3.11).

Figure 3.11 Cognitive Reading II Strategies

- Determining Importance – Identify the essential information
- Generating Questions – Engage in inquiries about the who, what, when, where, and why
- Monitoring and Clarifying – Check comprehension and understanding

Determining importance, generating questions, and self-monitoring are all cognitive reading strategies that make a difference in strengthening comprehension (Soto-Hinman & Hetzel, 2009). Young students should be encouraged to evaluate, question, and check their understanding as they read.

How the Cognitive Reading II Interventions Work

These three strategies help readers as they go through the reading process. Students should be continually thinking about their reading so that they can effectively understand the message being conveyed by the author.

Determine Importance

Students often read information without identifying the most important parts of their reading. Letting K-2 students know that information has an order of importance is extremely helpful to them. Our young students will face a future world where they will be inundated with data and information. As our students develop their ability to highlight the most important information they will get the most out of what they read.

- Who is the main character or what is the main message of the reading?
- Think about the top three ideas the author is emphasizing.
- What information supports the main idea?
- What information may be least important?

Generating Questions

Asking effective questions of our students can model good questioning for our students. There are a multitude of questions a young reader could ask as they read. The following list provides some very good questions that support the cognitive reading strategies (adapted from King, 1990).

- How are _____ and _____ alike?
- What do you think would happen if _________?
- In what way is _____ related to ______ ?
- How does _____ affect _____?
- Compare _____ and _____ with regard to _____.
- What do you think causes __________?
- Which one is the best _____ and why?
- What are some possible solutions for the problem of _____?
- What do I (you) still not understand about . . .?

Monitor and Clarify

Young students often spend a lot of their cognitive energy on decoding as they read. When students take the time to monitor and clarify their understanding as they read, they remember and retain more information.

- What does this word mean?
- Is this information making sense or do I need to reread?
- Wait, what's going on here, should I slow down?
- What text clues are helping me understand this?

K and ELL Scaffolding:

Because young students are just learning to read, as you read aloud to them check their ability to assign importance to information, ask questions, and monitor reading.

Cognitive Reading II Progress Monitoring

Again, it should be noted that cognitive reading strategies are important internal processes that need to be expressed to determine if they are working effectively. Create opportunities for students to evaluate the importance of the information they read, generate their own questions, and monitor their reading comprehension.

- Ask students to write down the main idea and key supporting information.
- Ask students to write down at least three questions that they generate in their minds as they read.
- Ask students to share which parts of the reading material they needed to slow down and think about more thoroughly.

Reading Intervention #5: Connecting with Text

It is important to make opportunities to connect with books and text that students read in deep meaningful ways.

Grimes (2006, p. 23) notes that connecting with text includes "readers using what they know to understand what they read by relating the text to their personal experiences, prior experiences with other texts, and their knowledge of world events and history."

Quality questions will help students delve deeper into their own understanding and they will make more meaning from the message conveyed by the author of the text.

What the Connecting with Text Intervention Looks Like

This strategy uses questions to get students to interact and connect with the text in ways that increase their knowledge (adapted from Johnson, 2009) (Figure 3.12 on pages 72 and 73).

Consider the following example from a 2nd grader's responses to informational text about planting a garden.

> When growing a garden seeds should be planted in the dirt. Seeds should be placed several inches under the ground and spaced several inches apart to give them plenty of room to grow. They need sunlight and water to grow. When seeds grow they send out a green sprout. Soon the green sprout will become a plant. Plants in a garden grow fruit or become vegetables. Eating fresh fruit and vegetables from a garden tastes good.

Figure 3.12 Connecting with Text

Connecting Text to Self

1. Ask students to make a connection from information in the text to something they already know.
   - "What do you already know that is related to the information in the text?"
   - "Does this remind you of anything in your life?"
2. Ask students to make a connection from someone in the text to something they already know.
   - "Who do you know that is similar to this character in the story?"
   - "How is this similar to your own life?"
3. Ask students if something in the text reminds them of any of their previous experiences.
   - "Has something like this ever happened to you?"
   - "What were your feelings when you read this?"

Connecting Text to Text

1. Ask students to make connections from the information in the text to other texts they have read.
   - "What other information have we studied that connects to the concepts in our text?"
   - "Have you read about something like this before?"
2. Ask students to make connections from characters in the story to other characters in other books.
   - "What other characters from stories you have read faced similar experiences or reacted in similar ways?"
   - "What does this remind you of in another book you've read?"
3. Ask students to make connections between the ways the author organized the patterns of text to other similar text patterns.
   - "What is the pattern of the text?"
   - "What other books do you know follow a similar pattern?"

Connecting Text to World

1. Ask students to make connections between the information in the text and some real world applications of the information.
   - "What does this remind me of in the real world?"
   - "How is this text similar to things that happen me in life?"

Figure 3.12 **Connecting with Text *(continued)***

2. Ask students to make connections between the experiences in the text and situations the student may face in the future.
   - "What can you learn from this character that will help you succeed in your future?"
   - "How is this different from things that happen in the real world?"
3. Ask students to make connections between the way the text is constructed by the author and how they structure their writing.
   - How has the author constructed the text?"
   - "What can you learn about good writing from the way this author structured the text?"

Connection Text-to-Self: I like strawberries, but I don't like broccoli.

Connection Text-to-Text: In "Jack and the Beanstalk" the magic seeds grow into a really tall bean stalk.

Connection Text-to-World: My mom buys watermelons from the market, and she lets me pick out the one I like.

How the Connecting with Text Intervention Works

1. **Prepare to read a story to students** by introducing the cover, title, and author.
2. **Read a story to students.** Read something engaging and interesting to the students.
3. **Ask students to make connections from the book to themselves.** Use the questions provided above.
4. **Ask students to make connections from the book to another book.** Use the questions provided above.
5. **Ask students to make connections from the book to the real life world**. Use the questions provided above.
6. **Invite students to share their answers with their neighbor** in a turn and talk.
7. **Invite students to then summarize the story,** if you want to work on writing skills.

> **K and ELL Scaffolding:**
>
> The connection with the text may be with the pictures in the text and how students connect the pictures to self, other text, and the real world.

Why the Connecting with Text Intervention Works

This intervention works because students benefit when they extend their thoughts about things they read. As students make connections to themselves they will personalize their reading. As they make connections to other books they will recognize relationships to other things that they are learning. As they think of examples of the real world they will discover more relevance in their reading. When students see that the books they read connect to their lives they will see the benefits of reading beyond just being entertained.

Connecting with Text Progress Monitoring

- If students are unable to connect a book to self, they should work with a partner to help generate ideas.
- If students have difficulty connecting a book to another book you can read two books in a row and ask the students to make comparisons.
- If students are unable to come up with examples in real-world situations, they may need more background knowledge.

Summing It Up

Students who struggle to read most often struggle in all of the core classes. Building a reader's confidence may make the biggest benefits in student confidence as a learner. Picture Word Inductive model is a very student-centered intervention that can be used with the whole group or in small-group instruction. This intervention effectively scaffolds student understanding by developing word knowledge and organizing information by classifying words into group patterns. Inferring is the number one way that students learn new words and build the vocabulary to read at grade level. Comparing texts will help students see the big differences between informative and narrative text. Word cards like the picture inductive model can add a wide variety of targeted vocabulary words to your students' vocabularies. And finally, as students learn to connect books to self, other books, and real-life situations, then they will extend and expand their understanding. As students increase their reading skills they will benefit throughout their school career.

Reflection

1. What reading strategies do your students need to develop to become better readers?
2. What types of inferences do your students readily know and explicitly use?
3. What kinds of connections do your students make as they read?

4

K-2 Intervention Math Strategies

"Arithmetic is numbers you squeeze
from your head to your hand
to your pencil to your paper
till you get the answer."

—Carl Sandberg

Every December, Luis misses school to visit all of his family in Mexico for a month around Christmas. Sometimes this trip extends longer than a month. This annual trip is so much fun, as he gets to see his cousins and there is always plenty to eat. His favorite activity is the piñata. Luis misses a lot of school. Luis does well in class when he is there, yet he misses his studies frequently throughout the school year. Whenever his little brother is sick, he is the one to stay home and take care of him. He likes to help his little brother and his family, yet he also knows that he gets behind in school. Luis is a frequently absent student.

Effective Interventions are Student-Centered

Effective interventions by their nature should be learner-centered. They should target the specific needs of the learners. Effective interventions should focus on what the learner brings to the process of constructing knowledge. Students may lack the resources to effectively construct learning—namely they may lack prior knowledge, academic language, critical thinking skills, and so on. Effective interventions work to achieve content objectives. At the same time interventions need to help students develop and achieve learning objectives. Achieving the learning objectives in many ways are more important in interventions than achieving the content objectives. Because once students develop the learning skills and strategies of a confident student, then they will be able to be more successful in all content areas.

Constructing Math Concepts

Most mathematics textbooks for K-2 students do a good job of outlining the basic computational skills and procedural skills. The ability to compute basic numbers is a vital foundational skill. Students need to learn procedures that they can execute fluently and with automaticity. Yet textbooks often emphasize rote procedures at the expense of conceptual understanding. For example, students who lack a proficient conceptual understanding often see math as a series of unrelated procedures or mechanical formulas. Students who only engage in limited procedural skills often fail to comprehend the mathematics principles behind the procedures. For example, students who are only taught a rote procedure for counting often believe that counting must be done from only left to right while assigning each item a number in succession. Better than this rote procedure, would be for K-2 students to understand the concept that items can be counted in any order as long as items are only counted once (Gelman & Meck, 1983). A complete conceptual understanding of counting provides students the key fundamentals for using more sophisticated strategies to add, subtract, and solve basic equations (Geary, Bow-Thomas, & Yao, 1992). If students learn how math concepts connect together, they will be better prepared to understand math more robustly. Chapin & Johnson (1997, p. 34) note:

> At the heart of mathematics is reasoning. One cannot do mathematics without reasoning.... Teachers need to provide their students with many opportunities to reason through their solutions, conjectures, and thinking processes. Opportunities in which very young students...make distinctions between irrelevant and relevant information or attributes, and justify relationships between sets can contribute to their ability to reason logically.

Where struggling students need additional support is in the underlying conceptual understanding strategies and problem-solving strategies. In this chapter we will look at intervention strategies that work for students who struggle early. A challenge with many math curricula is they use textbooks that are strong in procedural skills. It is important for students to become fluent in procedural skills. At the same time, students need to understand the primary conceptual connections and strategies that will make sure that the students truly grasp the mathematics concepts behind the procedures.

Conceptual Understanding

Conceptual understanding is important at all levels of study. For example, during the elementary grades students should understand that:

- One way of thinking about multiplication is as repeated addition.
- One interpretation of fractions is as parts of a whole.
- Measurement of distances is fundamentally different from measurement of area.
- A larger sample generally provides more reliable information about the probability of an event than does a smaller sample.

The strategies emphasized in this chapter may seem quite basic, yet they are concepts that young students need to be able to understand and execute in addition to basic computation. The five key interventions in this chapter are designed to make sure students have the means to more thoroughly and more explicitly learn the conceptual strategies for understanding math in a robust fashion. Students who develop a strong conceptual understanding in conjunction with procedural skills view mathematics as a way to create multiple solutions within the real world.

Mathematics Intervention #1: Multiple Representations

Math concepts can be challenging for some students to grasp. Truly understanding the concepts of math requires a fluent use of the language of numbers and symbols. The language of math is a language of symbols, signs, and numbers. Math is the language of creating—it is the language of how the universe is constructed. As young students begin the process of understanding and making meaning out of these numbers and symbols, it is important that they develop something commonly called "number sense." Gersten and Chard (1999) define number sense as a student's "fluidity and flexibility" with numbers. These authors note that students who are developing number sense should be able to understand numbers in relationship to:

- **Magnitude or size** – "Is eight closer to ten or one hundred?"
- **Comparison** – "Which number is larger: 9 or 23?"
- **Mental Computation** – "101 + 102 is 3 more than 200"
- **Estimations** – "30 – 8 is close to 20"

Number sense helps students understand real world quantities, mathematical symbols, and expressions. A proven intervention for effectively developing number sense is the Multiple Representations strategy that helps students recognize that numbers can be expressed in a number of written and visual forms. Representations work effectively with young students and they work with older students who are struggling in math. As students learn to use multiple representations in identifying numbers and concepts, they will strengthen their number sense and develop a more robust understanding of math. Shiro (1997, p. 11) notes several benefits of multiple representations.

> If children can use several different symbol systems (whether words, equations, diagrams, or manipulations of concrete materials) to describe the meaning of an insight they have into a mathematical or real-world experience, each symbol system will give them a different perspective on that insight, and the multiple perspectives can increase and enrich their understanding.

He continues by adding:

> The multiple representations of an idea can also allow children to better communicate their ideas to others, and can provide others with multiple ways of understanding and commenting on children's ideas.

Multiple representations can be used on the most basic concepts like a single number, and it helps students recognize that mathematical processes can also be solved through multiple methods. There are a variety of ways that math equations can be solved. Often math textbooks only show one standard way of solving equations. When students are able to express the solution to an equation through multiple methods, then their confidence as a math learner grows. In addition, they increase their ability to understand the math processes involved in the solution.

What the Multiple Representations Intervention Looks Like

Multiple Representations is an intervention that expands students' sense of how numbers and symbols can be recognized and used. At the heart of math is the use of symbols. A powerful factor for math is that the language is consistent across countries and cultures. For example, 2 + 2 = ? means the same thing around the world to different cultures. Yet, numbers and processes can be expressed in multiple ways. Take for example the number eight. This number seems simple enough, yet how many ways can a student express this number in their mind or on paper?

Ways to represent the number eight.

1. 8
2. 7 + 1
3. * * * * * * * * *
4. 10 – 2
5. 16/2
6. VIII
7. 2 × 4
8. 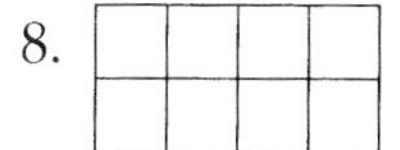

Students benefit tremendously from being able to represent a number or concept in their mind in multiple ways. The idea of adding, subtracting, multiplying, drawing, counting, or symbolizing a number concept provides tremendous flexibility to students as they use numbers to solve equations. In addition to the flexibility in understanding, multiple representations also help students with fluency as they look to solve equations.

How the Multiple Representations Intervention Works

This strategy works on several levels.

1. **Work with the whole class or pull a small group of students together** to work on a basic math concept (like the number 9).
2. **Ask students to think quietly of as many ways that they can write** or represent the number 9.

3. **Ask each student to share a different representation** until the students run out of answers.
4. **Add any other representations** to the list that you believe could be added.
5. **Ask students to do a simple page of math equations** with different representation for each number (see the sample assignment below).
6. **Review the math equations** to see if students are grasping the benefits.
7. **Discuss what the students learned** by looking at multiple representations for numbers.
8. To be really challenging, **ask students to write equations for their peers using mixed representation** numbers. This will build confidence and allow students to be creative with math.

You can use the sample assignment below to help students see that simple numbers and equations can be expressed through multiple representations (Figure 4.1).

Why the Multiple Representations Intervention Works

Multiple Representations works because it develops a student's number sense. The rather elusive concept of number sense is all about students developing fluency in working with numbers and basic math facts, while also developing automaticity. Research shows that expert learners view

Figure 4.1 **Multiple Representations Sample Assignment**

1. 1 + 1 = _____
2. *** + ** = _____
3. II + II = ______
4. 2 + III = _______
5. **** + 5 = ______
6. Two plus six equals _____
7. 4 plus 3 equals ______
8. Six + II = _______
9. _____ + 2 = 4
10. Three + ____ = six

information very differently than novice learners. Experts see things from multiple perspectives quickly and they can represent ideas and concepts in multiple ways. When students see math from multiple perspectives and they can do math processes in multiple ways, then they will see math as more than just a simple set of rote processes that have little meaning. They will start to see math as a method for creating solutions through multiple means.

Progress Monitoring for Multiple Representations

Students who struggle with Multiple Representations typically struggle with number sense. The speed at which they do math and the level at which the math sinks in is quite differently for students who struggle with number sense. You can check student progress for multiple representations in the following ways:

1. Engage students in the use of multiple representations to check on K-2 students' basic development of number sense.
2. The teacher will need to do this consistently for at least 12 weeks.
3. Once students can grasp math concepts or execute math equations fluently and with automaticity, it is time to move on to more challenging math equations using multiple representations.

As students become fluent in using multiple representations to understand math symbols and processes they will see math in much richer contexts and develop deeper meaning of math.

Mathematics Intervention #2: Tangible Manipulatives

Math is abstract and theoretical until it is brought into a real-world context. Math can explain processes that govern the smallest of molecules to the largest objects in the universe. Math can explain the inner workings of things, yet math may be best understood by young students by expressing math concepts in tangible terms. Many students in kindergarten, first grade, and second grade are kinesthetic learners who benefit by touching, interacting with, and manipulating objects that represent math concepts. Minton (2007, p. 83) emphasizes the benefits of tangible manipulatives:

> Manipulatives help move students from a concrete understanding to a more abstract understanding.

While Muschla, Muschla, and Muschla (2010, p. 253) note the variety of uses of tangible manipulatives:

> You can use manipulatives to model concepts, show relationships, and foster students' imagination and visual thinking.

Tangible manipulatives help students connect with basic math and they can concretely interact with the math to develop greater understanding. Tangible manipulatives and visual representations should be used as long as needed by students in the early grades (Checkley, 1999).

What the Tangible Manipulatives Intervention Looks Like

This strategy allows students to take abstract concepts like math symbols and numbers and interact with the concepts in tangible ways. The idea of abstraction can be very difficult for many K-2 students to grasp. As students start to see that abstract math concepts can be manipulated in tangible terms, then they will be able to learn more quickly and catch up to their peers. Most anything can be used as a tangible manipulative as students interact with math:

1. Popsicle Sticks
2. Connector Cubes
3. Chips
4. Beans
5. Cups
6. Students

For a creative teacher, the list is literally endless of the number of things that students can use to count or represent math concepts. For example, a teacher could add the number of students with red shirts (5) to the number of students with red shoes (2) to tangibly represent basic processes of addition. As students engage as a tangible manipulative or with tangible manipulatives they will be able to see how the math makes sense more clearly.

How the Tangible Manipulatives Intervention Works

You will notice that these strategies work with struggling students, yet they also work well with all students.

Let's look at how using tangible manipulatives can help students learn

1. **Place students in pairs,** matching up high and low learners.
2. **Give students several simple math equations to solve.** They can work in pairs.

3. **Provide students with tangible manipulatives** that they can use to manipulate and represent the math.
4. **Model for students** how the tangible manipulatives can help students better understand the math.
5. **Ask students to show their math solutions** to the teacher or their partner using the tangible manipulatives.
6. After finishing the assigned math equations, **students can rotate around the room** showing the math through the tangible manipulatives to different classmates.

Young students enjoy showcasing their new found understanding to themselves and for their peers.

Why the Tangible Manipulatives Intervention Works

As students are able to handle tangible items that serve as identifiers for abstract concepts, they can more readily grasp the concepts in real terms. It is fairly obvious that tangible manipulatives help kinesthetic learners understand math in conrete ways. At the same time, visual learners will be able to see how the math works as they use tangible manipulatives. As students talk about the process they are demonstrating, then auditory learners will also benefit.

Progress Monitoring for Tangible Manipulatives

Because this intervention works very well with struggling students and all students alike, this is an ideal strategy for Tier I classroom interventions. It also works well in Tier II small-group interventions.

- Tangible manipulatives provide an easily identifiable method for seeing if students are fully grasping concepts as they explain their understanding.
- Observe students' use of tangible manipulatives to see how many ways they can represent concepts.
- Learning basic counting concepts, sets, and benchmark numbers like 5 and 10 with tangible manipulatives works well, so ask students to showcase their learning with the tangible manipulatives often.

Make sure that you have a variety of tangible manipulatives that students can use every day to connect their learning to abstract math concepts.

Mathematics Intervention #3: Math Dialogue

The more students engage with math, the more they will process the ideas and learn the concepts of math. Students can engage with math by doing lots of practice, learning routine procedures, considering things from multiple perspectives, manipulating math objects, and discussing the ideas out loud with others to clarify understanding. Making meaning is a negotiated process, and in addition to negotiating with the teacher and the author of books, students seem to most enjoy negotiating meaning with their peers. Students need plenty of opportunities to discuss math concepts and engage in math dialogue. The more students discuss their growing understanding of math out loud, the quicker the ideas in their mind will be clarified and concepts will be organized and reinforced. This dialogue process is extremely "foundational to children's learning" (Wood & Turner-Vorbeck, 2001, p. 186).

1. **Ideas are important, no matter whose ideas they are.** Students can have their own ideas and share them with others. Similarly, they need to understand that they can also learn from the ideas that others have formulated.
2. **Ideas must be shared with others in the class**. Correspondingly, each student must respect the ideas of others and try to evaluate and make sense of them.
3. **Trust must be established** with an understanding that it is okay to make mistakes. Students must come to realize that errors are an opportunity for growth as they are uncovered and explained.
4. **Students must come to understand that mathematics makes sense.** As a result of this simple truth, the correctness or validity of results resides in the mathematics itself. There is no need for the teacher or other authority to provide judgment of student answers.

Providing students with frequent (even daily) opportunities to engage in math dialogue will go a long ways to creating a "mathematical community of learners" (Thomas, 2009).

What the Math Dialogue Strategy Looks Like

Vygotsky (1979) emphasized that constructive learning is a social interaction. Students need opportunities to process their understanding by verbally presenting and sharing it with others. Discourse with peers provides a platform for students to present their ideas to others. Shared dialogue in mathematics allows students to insert themselves in the math conversation.

Restivo, van Bendegem, and Fischer (1993, p. 157) point out, "Engaging in genuine dialogue within mathematics classrooms quite naturally leads to dialogue about mathematics as an area of personal interest."

Ultimately, math is the language of creating in the physical universe. The authors of math textbooks chime in with their perspectives on learning math skills and concepts. Students also need to find their "math voice" so that they can grasp the concepts that will allow them to create and build. When students engage in math conversations with their peers, they are able to contribute to serious learning. Shared dialogue helps students understand solutions to math equations. The following four-step dialogue frames engage students:

1. Saying "I think that... because..."
2. Understanding "I listened to what X said and now I think..."
3. Reflecting "My understanding was/was not correct because..."
4. Rethinking "I now think that... because..."

As students value the ideas of others and openly share their thinking and understanding of math concepts and processes, they will see things from additional perspectives. Instead of math being a boring time of silently solving a long list of grueling equations, math can be an interactive, vibrant conversation of meaning and understanding. This intervention will go a long ways to creating a true community of math learners.

How the Math Dialogue Intervention Works

Most any math concept can be used as a basis for engaging your students in math dialogue. It is important to clearly know what your objective is when you assign a math concept to be explored and discussed by your students, so that you can help guide them to a productive result. The following steps will help support math dialogue in your classroom.

1. **Provide students with a challenging math concept** and outline some of the potential difficulties they may face.
2. **Place students in pairs** and assign them so that stronger math students are paired with students who need more help.
3. **Ask students to think, pair, and share** and as students think they can draw out their understanding of the concept, identify patterns, write down an example, think of a real-life situation, and so on.
4. **Give students time to negotiate meaning** and question each other or poke, prod, extend, and expand ideas.

5. **Place two pairs together in a group of four** and then ask the students to share with the other partners.
6. **Ask students to symbolize their negotiated understanding** through a picture, equation, example, written statement, and so on.
7. **Next, groups of four can share their understanding** with the entire class through a teacher directed discussion.
8. **Finally, make sure that concepts are challenged** and the students' thinking is accurate.

When students can justify their explanations verbally, then students have learned at a very high level and they will more easily retain this information for future use.

Why the Math Dialogue Intervention Works

On a daily basis, students should be given the opportunity to share their understanding of how the math is working with a peer. Matching students who have a good grasp of the math concepts with someone who is still developing their understanding is an efficient use of time and an effective method for strengthening conceptual comprehension. The challenge for students to articulate math concepts is in fact a meta-cognitive process that makes known to learners what they know and helps them identify areas where they may still lack understanding. Creating a community of math learners is an important step in changing students' perceptions of math. When math is a regular part of classroom conversations, then it becomes much more than just a long list of rote procedures to be executed silently. Second Language Learners who may know the math will benefit by trying to find words in English to express their understanding. Students are able to find their "math voice" as they engage in math dialogue. Students will be much more engaged as they negotiate with their peers and learn math together rather than in isolation.

Progress Monitoring for Math Dialogue

Listening to students engage in shared dialogue as they explain math concepts to one another is an extremely effective way to monitor a student's understanding of the concepts.

1. Give students who are struggling more time in small-group instruction.
2. Listen to students answers and help them if they get stuck.
3. Review students' explanations, examples, pictures, written outlines, and so on to see if math concepts are being understood at a rigorous level.

Consistently engaging students in shared dialogue about math concepts may be the number one activity that turns your classroom into a "mathematical community of learners."

Mathematics Intervention #4: Conceptual Models

New concepts in math should be presented in concrete, pictorial or representational, and abstract terms. Visualizing and representing mathematical concepts helps students solve math problems. Charlesworth (2004, p. 3) notes that conceptual models can help students understand basic math processes like addition and subtraction.

> As children enter the primary period (grades one through three), they apply these early basic concepts to help them understand more complex concepts in mathematics such as addition, subtraction, multiplication, division, and the use of standard units of measurement.

The following list adapted from Posamentier and Kulik (2009) provides a variety of ways to organize conceptual models in mathematics:

- Organizing the Data
- Creating a List
- Making a Table
- Intelligent Guessing and Testing
- Solving a Simpler, Equivalent Problem
- Acting It Out/Simulating the Action
- Working Backward
- Finding a Pattern
- Logical Reasoning
- Making a Drawing
- Adopting a Different Point of View

What the Conceptual Models Intervention Looks Like

As students learn new math concepts, models can assist students in more fully grasping key concepts. Models help represent the relationships between ideas. Identifying how ideas connect will help students fully conceptualize math. Models can be designed in a variety of ways. Most of the time we think of models being a physical representation of abstract ideas. Models in fact can be a visual or physical representation of a math concept. Van de Walle and Lovin (2006) identify several important steps when creating math models:

- Introduce new models by showing how they can represent the ideas for which they are intended.
- Allow students (in most instances) to select freely from available models to use in solving problems.
- Encourage the use of a model when you believe it would be helpful to a student having difficulty.

Mathematical concept models can be used to help students see multiple ways to understand mathematical process. For example, the concept of addition can be conceptually understood in more than one way. Addition and subtraction can be viewed as a join-separation process:

1. Initial	Six	
2. Change	subtract Two	$6 - 2 = 4$
3. Result	equals Four	

Join-separation equation processes have three quantities involved that can be expressed as:

Join: Initial Unknown	Bobby had some pennies. Jill gave Bobby 2 more pennies. Now Bobby has 8 pennies. How many pennies did Bobby have to begin with?
Join: Change Unknown	Bobby had 6 pennies. Jill gave him some more pennies. Now Bobby has 8 pennies. How many did Jill give him?
Join: Result Unknown	Bobby had 6 pennies. Jill gave Bobby 2 more pennies. How many pennies does Bobby now have?

Or you can look at addition/subtraction as a part-part-whole pattern:

1. Part
2. Part
3. Whole

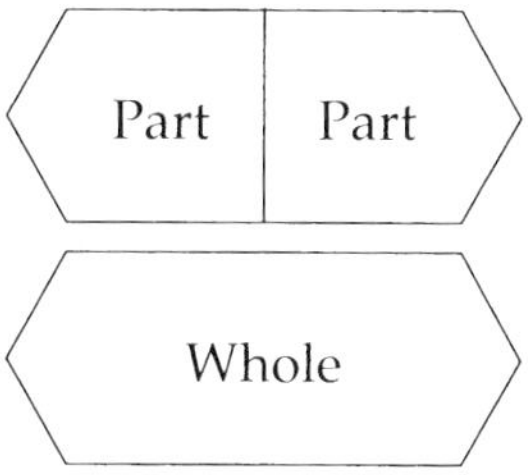

Part	Part
Whole	

Or you can look at addition/subtraction as a set-comparison pattern:

1. Large Set
2. Difference
3. Small Set

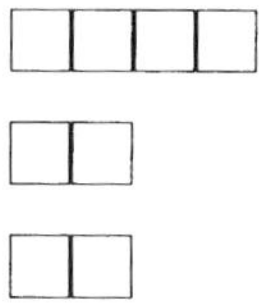

How the Conceptual Models Intervention Works

Students need a variety of models provided to them so that they can start to construct their own mental models for themselves. Einstein would create mental models to conceptualize how math operated within the universe. Visual models can be drawn by students.

Students should be given tangible resources to construct mental models. Toothpicks, connector cubes, can help students to create their own math visualizations (Figure 4.2).

Figure 4.2 **Math Visualizations**

Visual models

1. **Invite students to consider a concept they have been learning** like counting in tens.
2. **Asks students to draw a diagram** of how the concept works.
3. **Ask students to explain** how the model works to a peer.

Physical models

1. **Invite students to consider a concept they have been learning** like subtraction.
2. **Ask students to design a model** using tangible manipulatives of how the math concept works.
3. **Ask students to explain** how the model works to a peer.

The key is for students to think through and cognitively consider various methods that would effectively showcase their understanding.

Why the Conceptual Models Intervention Works

This strategy works particularly well because it engages meta-cognitive processes within the students. Conceptual modeling helps students start to see math in pictures and they see math as a way of creating solutions (Lesh, Post, & Behr, 1987). Helping students design visual and physical models will build their confidence and they will be able to showcase their math to parents, peers, and their teacher. The conceptual model gives students a full understanding of math with deeper meaning and a more complete comprehension.

Progress Monitoring for Conceptual Models

Young students may need a lot of support and help using conceptual models. If it takes students a while to get the hang of conceptual models it is okay. Work with these struggling students together in a small group. Check to see if they can:

- Identify all of the component parts.
- Identify all of the important processes.
- Put all of the parts and processes together in a comprehensive conceptual model.

Young students who frame math as conceptual models will continue to be motivated to do math instead of discard math as a bunch of silly rules and procedures with little meaning.

Mathematics Intervention #5: Problem Solving

We face challenges that may be defined as problems it seems every day. As we go through life we are asked to solve a myriad of situations or problems. Becoming a good problem solver is important. Wall and Posamentier (2006, p. 82) point out:

> Problem solving is the cornerstone of school mathematics. Students who can efficiently and accurately multiply but who cannot identify situations that call for multiplication are not well prepared. Students who can both develop and carry out a plan to solve a

mathematical problem are exhibiting knowledge that is much deeper and more useful than simply carrying out a computation.

Developing our students' problem solving disposition and abilities helps them better understand math. Bell and Bell (2007) note,"Children develop a better understanding of various mathematical processes when asked to think and strategize, rather than when they are merely asked to repeat the steps of a process."

As we help students approach math as a problem solving scenario, then students will be more engaged and utilize their native abilities to recognize patterns and create solutions.

What the Problem Solving Intervention Looks Like

We are born problem solvers. We naturally look for ways to improve a situation. Coming from a problem solving perspective can help students see the potential relevancy of the equation. Yet, becoming good problem solvers may take practice. In addition to learning the basic concepts of number sense, students need to develop a problem solving disposition. Posamentier and Krulik (2009, p. 2) define problem solving in these terms: "A problem is a situation that confronts the learner, that requires resolution, and for which the path to the answer is not immediately known." They continue: "Although some student intuitively may be good problem solvers, most of our students must be taught how to think, how to reason, and how to problem solve."

Students need to be given a real-world context to help see where the abstract math meets a concrete, real-life situation or equation. If students learn math as only a set of rote procedures and limited rules, they will struggle to see math as an engaging process that can help in learning mathematical principles as well approach challenges in life.

The following list provides a variety of ways that numbers and math principles can be used to engage in problem solving.

- **One-to-one correspondence:** pass out one to each child, put pegs in a board, matching a block to a number.
- **Counting:** pennies, number of jelly beans, and number of rocks in a collection.
- **Classifying:** placing square shapes in one pile and triangles in another pile.
- **Measuring:** pouring rice into other containers and seeing amounts.

As students master solving problems in the preceding areas, they can then progress on to more challenging situations.

How the Problem Solving Intervention Works

Problem solving must require justifications and explanations for answers and methods.

1. **Provide students with a simple task**—For example, finding out how many boys and girls want to buy lunch from the cafeteria.
2. **Ask students how to frame a potential solution**—Work with students to properly design the equation that will arrive at a solution.
3. **Provide students with time to explain** their solution to a peer.
4. **Ask students to justify** why they chose to frame the problem and arrive at a solution.
5. **Ask students to think of other areas**—This problem-solving skill may be applied in real-world situations.

Why the Problem Solving Intervention Works

Problem solving helps students exhibit their natural curiosity, intelligence, and flexibility in facing challenging situations. When students see math as a solution to a problem or challenge, they see the relevancy of the math more clearly. Students want to solve real-world issues, and math is an effective way of developing the skills to solve challenges. Students engage in problem solving more than when they are merely computing rote equation upon rote equation.

Progress Monitoring for Problem Solving

The key is for real-world situations to be expressed. Students should create their own story problems or context solutions for their peers.

- Encourage students to ask lots of questions as they develop a mathematical disposition.
- Listen to the questions students ask themselves or each other as they engage in problem solving.
- Students should justify and explain their thinking and the processes they explored, so review written or spoken justifications.

When students create their own problem-solving tasks for their peers they should also be asked to justify or explain why they set the problem in the context that they did.

Summing It Up

Teaching math to young students can be challenging. Many early math textbooks and manuals only present math as rote procedures to be memorized with little student understanding. While students need to learn basic procedures, it is important that they recognize the depth and variety of ways math can be understood. Multiple representations helps kindergartner and other young students see that math can be approached from multiple perspectives and there are multiple ways to arrive at a solution. Tangible manipulatives provide students a means for grabbing a hold of abstract math concepts so that the students can put these principles into real-world terms. Shared dialogue helps students process through their own understanding and they can negotiate meaning with their peers. Concept models can be very basic for young learners, yet they help students see the multiple parts that go into math processes. Problem solving is a disposition or attitude that helps students see that math can be used to arrive at real world solutions. If math is only presented as rote procedures, it can be viewed as senseless pages of numbers. When students see multiple ways to represent mathematical concepts and can tangibly express their understanding, then math will have meaning and purpose.

Reflection

1. In what ways do you make math engaging for your students?
2. Do all of your students have a strong understanding of number sense?

5

K-2 Intervention Speaking Strategies

"Words mean more than what is
put down on paper.
It takes the human voice
to infuse them with shades
of deeper meaning."

—Maya Angelou

Nataliya is a naturally shy girl that seemed drawn into a shell. Even though she seems to say a lot with her bright blue eyes, she says very little with her words. She immigrated to Seattle with her parents and two older sisters four years ago from Kiev. Her family speaks Ukrainian in her home. At school, she looks like the other students in her class, yet the language, customs, and conservative culture of her family are so very different from her peers. She never speaks at school unless someone directly speaks to her. She avoids conversations as best she can. She likes school, yet it seems that many of the words are too difficult for her to understand. Her parents are very supportive of education and they make sure that she completes her homework. Even though she does her school work the best she can, her language skills seem to be progressing slowly. At lunch time she speaks in Russian or Ukrainian to her few friends. Avoiding conversations in English seems to be taking a toll on her ability to learn the words for learning. Nataliya is a second-language learner and she is a struggling student.

Speaking strategies as a response to intervention makes sense on many levels. Students receive tremendous benefits when they receive multiple opportunities to speak in class. Through the process of speaking in class, our students reveal their thinking processes to others as well as themselves. When they speak, they also more fully engage with classroom content and new information. They benefit as they negotiate meaning and participate in content-area conversations. Meaning is a negotiated process. So, for students to make meaning they need to be given situations where they can discuss their understanding and thinking. As instructional leaders in the classroom, we should integrate a variety speaking strategies in large-group, small-group, and paired-partner activities. The strategies should be able to meet the needs of all students and provide support for socioeconomically disadvantaged students, English Language Learners, and struggling learners.

Cracks, Gaps, and Chasms

The achievement gap and the growing dropout rate reveal the significant challenges many of our students face. Cracks in the foundation of students' language can appear before students even enter kindergarten. If these cracks in a student's language foundation are left alone, they often widen over time. Eventually if these language cracks are ignored, they can lead to gaps in literacy and can eventually overwhelm our students' ability to learn. Simply stated, the language cracks that appear in kindergarten and the early elementary grades become significant literacy gaps in the upper elementary grades. In time, as students transition from middle school to high school learning chasms begin to appear that swallow up entire groups of socioeconomically disadvantaged students, struggling readers, and second-language learners.

Language Cracks – Cracks begin to show up in the learning foundation as students enter school.

Literacy Gaps – Over time, gaps begin to appear in the framework that holds learning together.

Learning Chasms – Eventually, chasms begin engulfing large groups of students (especially the poor, minorities, and struggling readers), and they face growing dropout rates.

Figure 5.1 **Identifying How Achievement Gaps Affect Learning**

Language Cracks
↓
Literacy Gaps
↓
Learning Chasms

Unless the cracks are systematically addressed using targeted interventions, the cracks will become gaps and eventually chasms. For example, I (Eli) served as an administrator at an inner-city high school in a large metropolitan area. We had 800 freshmen enter our campus each year. Four years later we graduated approximately 400 seniors. The result was effectively a 50% dropout rate. What happened to these students? Where did they go? They slip into the cracks, get wedged into the gaps, and eventually fall into the chasms of our educational system. The language cracks that start to appear in elementary become literacy gaps by the time many of these students reach middle school, and by the time these students reach high school the cracks and gaps become learning chasms. These chasms are swallowing entire groups of students, and they are dropping out in monumental numbers (Figure 5.1).

Kids seem better able to succeed with the language and learning of the streets than with the language and learning of school. Is it any wonder that so many of our students are dropping out in such large numbers? We need effective strategies that work to fill the cracks, gaps, and chasms that show up in our students' learning.

Speaking and Socioeconomic Disadvantaged (Title I) Students

Students from poverty typically have fewer opportunities to engage in conversations within their home. They need classroom activities that will direct them in many different ways to produce language and express their understanding. The process of improving thinking improves as students express their ideas verbally. Those students from homes in poverty are spoken to less when compared to their wealthier peers. In fact, they receive one half as many speaking opportunities than middle class students, and less than one third the speaking opportunities compared to their high socioeconomic peers (Hart & Risley, 2003).

Speaking and English Language Learners (ELLs)

Second-language learners often have few chances to speak English outside of school. The time spent speaking in school is critical for these students to engage in meaningful conversations. As our students receive structured activities to speak in school, they can make tremendous progress in their ability to communicate with others. Without specific speaking strategies in math, social studies, science, and language arts, ELLs can get further and further behind their peers. Frequent speaking opportunities can benefit all students' understanding and meaning, yet ELLs will benefit even more because of the limited chances to speak the language outside the classroom. Research shows that in many classrooms English Language Learners are only asked to produce language and speak in class 4% of the time, and only 2% of this student speaking is academically focused.

Speaking and Struggling Learners (RTI)

Students who struggle at school typically are reluctant to engage in activities that will reveal their thinking. These students need carefully scaffolded activities that are sensitive to their reluctance, yet provide a variety of strategies for students to speak and produce language.

It is important to match up struggling learners with positive and supportive partners who can effectively model the strategies while also patiently encouraging their partner.

Speaking Intervention #1: Sound Muncher

The Sound Muncher intervention is for K-2 students who need extra attention developing their phonemic awareness and pronunciation of sounds in the English language. Young students absolutely love this kinesthetic strategy that allows them to receive immediate feedback that recognizes their correct responses as they properly produce the 43 sounds in the English language. Bray (2007, p. 5) notes:

> Proficiency in sound isolation leads to sound blending, the ability to combine individual sounds to make words. Higher levels of phonemic awareness include sound segmentation—the ability to identify the individual sounds in words—and sound manipulation, which allows children to create new words by deleting, adding, or substituting sounds in words.

Figure 5.2 Sound Muncher

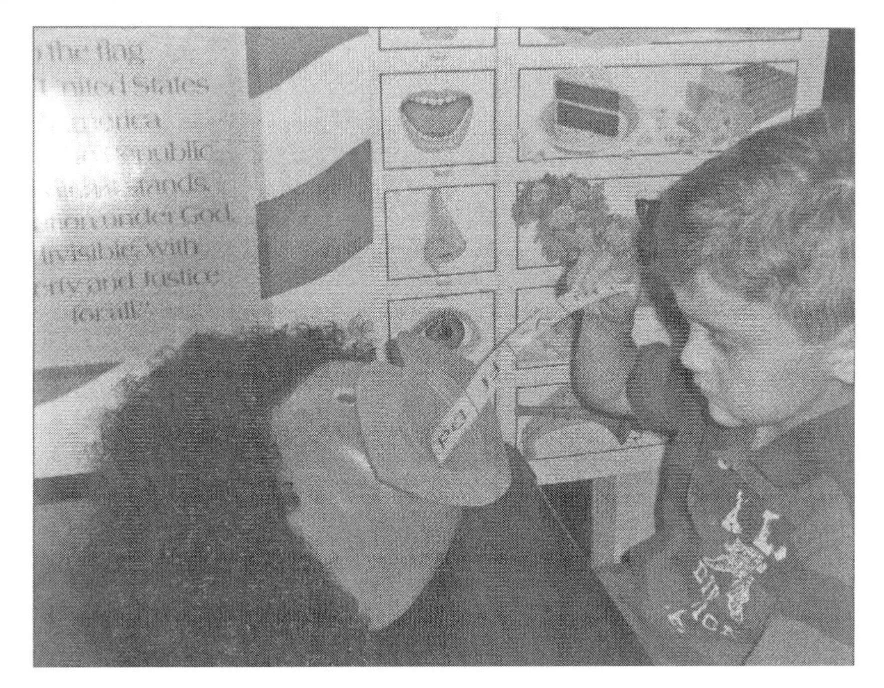

The Sound Muncher is really just a well-disguised mini-garbage can or other receptacle dressed up to serve as the mouth ready to digest cards. Sound Munchers have eyes, and a nice wig to outfit them—earrings are optional. Students enjoy placing cards that they have answered correctly into the Sound Muncher.

What the Sound Muncher Intervention Looks Like

The Sound Muncher is an intervention for K-2 students who need extra attention developing their phonemic awareness and pronunciation of sounds in the English language.

- Basic single consonant sounds
- Long vowel sounds
- Short vowel sounds
- Combination sounds

Consonant sounds come in two primary forms: continuous sound phonemes and stop sound phonemes (Figures 5.3, 5.4, and 5.5 on pages 102 and 103).

Figure 5.3 **Consonant Phenomes and Sounds**

Consonant Phonemes	Sounds
Continuous Sound	/f/, /l/, /m/, /n/, /r/, /s/, /v/, /w/, /y/, /z/
Stop Sounds	/b/, /d/, /g/, /h/, /j/, /k/, /qu or kw/, /t/ /x or ks/

Figure 5.4 **Phonics Elements and Sound/Spelling Categories**

Phonics Elements	Sound/Spelling Categories
Single Consonants	(b, c, d, f, g, h, j, k, l, m, n, p, q, r, t, v, w, x, y, z)
Short Vowels	(a, e, i, o, u)
Long Vowels	(a, e, i, o, u) and long vowels with silent e (a_e, e_e, i_e, o_e, u_e) as in (ate, scene, ice, hope, cute)
Consonant Blends	Two consonant letters that work together in a word and each retains its sound when blended (i.e., fl, gr, sp, mp)
Consonant Digraphs	Two consonants that together make a single sound. (i.e., sh, th, wh) as in (shut, that, when)
R-Controlled Vowels	The vowel before the r is treated as a single sound. (i.e., er, ir, ur, ar, or, air) as in (her, fir, fur, for, pair)
Vowel Digraphs	These combinations make one sound. (i.e., ai in wait, ee in feet, ie in tie, oa in boat) Variant digraphs are aw in saw, au in taut, oo in tool, oo in good)
Dipthongs	The blending of vowel sounds in one syllable. (i.e., oi in soil, oy in toy, ow in cow, ou in cloud)
Schwa	The vowel sound sometimes heard in an unstressed syllable (i.e., like the a in about, or the e in taken, or the i in pencil, or the u in supply, or the y in sibyl).

(Adapted from Wyle & Durrell, 1970)

How the Sound Muncher Intervention Works

The Sound Muncher strategy asks students to quickly produce the phoneme or sound associated with a letter or underlined letter in a word. For example students would be asked to correctly say the long vowel sound in "E" or the soft vowel sound in "bat."

Figure 5.5 **Common Rhyming Endings**

-ack	-ail	-ain	-ake	-ale
-ame	-an	-ank	-ap	-ash
-at	-ate	-aw	-ay	-eat
-ell	-est	-ice	-ick	-ide
-ight	-ill	-in	-ine	-ing
-ink	-ip	-ir	-ock	-oke
-op	-ore	-or	-uck	-ug
-ump	-unk			

1. **Model for students** the appropriate pronunciation of various *sounds* and their correlation with letters.
2. **Show a student a card** with a letter or combination of letters that represent a specific sound.
3. **Hand a card to a student** and invite the student to pronounce the sound that corresponds with each card.
4. **Students get to place the card in the Sound Muncher**, when they correctly pronounce the proper sound.
5. **If students are incorrect, the card is put down in a pile**, and the next student in the small group should be given the opportunity to see if they can pronounce the word correctly.
6. **Continue through the stack of cards;** sounds that students struggle pronouncing can be reviewed at the end of the stack.
7. **As students progress through letters**, they should then be given lists of words and be asked to pronounce the sounds made by each letter in the word.
8. **Again, as students pronounce all of the sounds correctly** the student can then place the card with the word on it into the Sound Muncher.

K and ELL Scaffolding:

Make sure that young students have lots of chances to say sounds they can produce successfully, so that they get to stick their cards into the Sound Muncher (consonants like b, p, k, and t are often easier to produce for new students).

Why the Sound Muncher Intervention Works

The Sound Muncher intervention works because the strategy is very engaging for young students. What are the rules? They love the chance to place their successful responses inside the Sound Muncher container after correctly pronouncing the right sounds. The strategy moves quickly and a lot of sounds can be covered in a short amount of time. K-2 students love the immediate feedback that lets them know if they are correct. The reward of placing their correct answers into the Sound Muncher is a small, yet tangible reward. If a student mispronounces a phonemic sound, then the next student is given the opportunity to pronounce the sound outlined on the card. This creates a need for students to pay attention as their peers are asked to pronounce letters, and they will benefit from their attentiveness. The Sound Muncher gives immediate recognition regarding the correctness of student responses. It builds confidence for students as they enjoy the reward of visibly placing their correct answer into the Sound Muncher's mouth.

Progress Monitoring for Sound Muncher

The teacher can move from student to student quickly as they go through a stack of phonemic awareness cards. As students master their basic sounds, then more challenging digraphs can be practiced. Eventually, students can practice blending and segmenting sounds during the small-group time.

- Listen to make sure students are enunciating the sounds and blends properly.
- Give students more time with this engaging strategy if they continue to struggle.
- Encourage students to listen to others so that they can start to identify proper sounds from others.

Sound Muncher makes the standard process of producing sounds a more engaging experience that students will enjoy because of its kinesthetic process and immediate reward.

Speaking Intervention #2: Fluency Phones

Students need ample opportunities to practice fluency so that they can improve their rate, accuracy, and prosody. Mash and Barkley (2006, p. 553) write that

...Interventions involving fluency were efficacious, with virtually all of the studies examined showing gains in response to instruction.

Wiley Blevins (2002, p. 23) notes the following areas for targeting fluency with quality interventions:

- Pinpoint particular areas of intervention and target instruction.
- Provide ample opportunities to read decodable text and high-interest easy reading.
- Establish regular practice where students can hear themselves speak.
- Provide many chances for reading and rereading passages.
- Set up a listening center where students practice with increasing length and complexity.

The fluency phone provides an engaging way for students to practice fluency while staying engaged and maintaining interest.

What Fluency Phones Look Like

Fluency Phones physically look like a phone. The total cost of each phone is approximately $3 dollars. The materials needed to make them can be found at most any local hardware store. The Fluency Phones are made of three white plumbing pipes: two 1½" corner pieces and a 2" straight piece. The three parts simple push together to create a phone that provides terrific amplification for students. The Fluency Phones are a great tool that all students can use in class at the same time. For a class of 30 students, at $3 a phone, the total would be around $90 to outfit the entire class. A serendipitous benefit of the fluency phones is the improvement of the courtesy and social graces indicative of age-appropriate conversations. Students can use Fluency Phones each day to practice reading while they hear the sounds they are producing while speaking (Figure 5.6 on page 106).

How the Fluency Phones Intervention Works

1. **Make phones for students** and provide one for every student.
2. **Teachers create packets where the phones are kept** and reading passages can be changed out by the teacher on a weekly basis. It is best if the dialogues reflect the reality of the classroom. It is a way to hook the students into the process because they like talking about each other and using real-time information keeps interest high.

Figure 5.6 **Fluency Phones**

3. **Students should get their fluency phones**, reading packets, and prepare to start talking by reviewing the content and the sounds they are expected to reproduce.
4. **Passage selected for students should have high interest** for the students.
5. **Circulate throughout the room** for at least five minutes and listen to students.
6. **Once students are reading on their own, select five students** to pull into a small group to conduct a variety of intervention strategies.
7. **Once students complete their daily reading** they can put their Fluency Phones and reading packets away before starting their May Dos and Must Dos list.

> **K and ELL Scaffolding:**
>
> While many of us have dabbled with Fluency Phones in the classroom, the consistent daily use of the phones by every student can help them progress quickly (Remember every student every day for at least 10 minutes of reading practice).

Why the Fluency Phones Intervention Works

This intervention works because students need frequent practice reading every day. The Fluency Phones are an excellent way to get students to read as an entire group while they still get to hear themselves. Fluency Phones appeal to auditory learners who can effectively hear themselves to make sure that they are pronouncing words properly. Fluency Phones also appeal to kinesthetic learners who get to hold the phone and talk into it. This is an extremely engaging intervention that students enjoy doing every day. The reading passages can be specifically selected for each student according to their abilities.

Progress Monitoring for Fluency Phones

Students can record their speed as they go through their passages and see their progress. The teacher can circulate around the room and listen to students to as they read into their Fluency Phones.

- Listen to students at the beginning of Fluency Phone time every day to hear how things are progressing.
- Track the fluency progress for students so they can see progress.
- Encourage effective expression, intonation, and feeling, so that students focus on more than their speed at decoding.

Pull students into small groups. Students can record their speed as they go through their passages and see their progress. Charting progress in a public manner, getting a written note of praise, or hearing from the teacher that improvements are being made are motivating and will prompt continued attempts to increase rate, prosody, or experimentation with language.

Speaking Intervention #3: Structured Discussions

Sentence frames provide students with a comfortable framework to structure their classroom conversations. Sentence frames are an ideal method for getting students started in the right direction. Sentence frames give students a common starting point, yet the students direct the end results the discussion will produce. Sentence frames or sentence starters can be used to introduce students to new academic language and connect content-area concepts.

Cooper, Kiger, and Au (2008, p. 97) point out, "A structured discussion is an excellent way to assess students' prior knowledge; responses and interactions will reveal their knowledge and their misconceptions." Sentence frames can scaffold for students the language to help them access core content in an efficient and effective manner. Structured discussion is an excellent way to listen to students' knowledge and understanding (Marzano, et. al. 2001).

What Structured Discussions Look Like

The following discourse practices benefit all students, as they increase many of the communication skills they are taught at home. For the many students who lack these skills, consistent practice will improve the quality of classroom dialogue. Students should practice these classroom-speaking skills daily. Students need to understand the function of language and the purposes that can be accomplished. Underscoring that listening and reading are passive skills, students need to know that their ability to use language is measured by their ability to communicate their ideas by speaking and writing. In English Language Assessments, students need to be able to communicate in school in the following ways (CELDT).

- Asking for Information – "Excuse me, what is a good word for ugly?"
- Making a Request – "May I have a drink of water?"
- Offering Assistance – "Would you like me to help?"
- Asking for Clarification – "Did you say turn right in the hallway?"
- Give an Explanation – "Teachers ask questions to make us think."

Figure 5.7 Structured Discourse Skills

- **Expressing an Opinion** – "I believe that" or "I think" or "In my opinion"
- **Paraphrasing** – "I hear you saying that your feelings got hurt.
- **Asking for Clarification** – "Let me check, did I hear you say?"
- **Soliciting a Response** – "Do you really believe in angels?"
- **Acknowledging others Ideas** – "What a great idea!"
- **Offering a Suggestion** – "I think Paco's has good food."
- **Agreeing** – "I really like that pink coat on you!"
- **Disagreeing politely** – "That is not my favorite color."

Figure 5.8 Structured Discussion Sentence Frames

Expressing an Opinion	**Reporting a Group's idea**
♦ I believe that . . .	♦ We agreed that . . .
♦ It is my opinion that . . .	♦ Our group believes that . . .
Paraphrasing	**Disagreeing Politely**
♦ In other words, you think . . .	♦ Another idea is . . .
♦ What I hear you saying is . . .	♦ I see it another way . . .
Soliciting a Response	**Offering a Suggestion**
♦ What do you think about . . .	♦ Maybe we should consider . . .
♦ Do you agree with . . .	♦ Here's something we might try . . .
Acknowledging Others Ideas	**Agreeing**
♦ My idea is similar to _______'s idea . . .	♦ That's an interesting idea, I agree that . . .
♦ I agree with _______ that . . .	♦ I hadn't thought of that, I like . . .

Students need plenty of support to engage in a structured discussion that keeps conversations on track. Sentence frames help get students started down the right track as they begin a structured discussion (Figure 5.8).

Students will benefit by beginning their conversations in class with sentence frames. The Structured Discussion Sentence Frames can be written on the backs of cards and help young students develop their classroom communication skills.

How the Structured Discussions Intervention Works

Teachers need to plan and organize classroom discussions, whether they are large-group or small-group discussions, in ways that structure the thinking and conversations in ways that will engage students in using academic language (Figure 5.9).

Kindergarten students may need to have the sentence read to them and work as an entire group as they practice structured discussion skills. Writing the sentence frames on cards for first- and second-grade students will help them significantly.

Figure 5.9 Structured Discussion Pairs

1. **Place students in pairs** that share a common performance level. Each pair should have a short paragraph of text that they read. Nonfiction and real-time excerpts using student names and the current machinations of their classroom lives.
2. **Model taking turns,** providing interpretive comments and receiving feedback responses.
3. **Ask the students to read** and have them make one significant statement to summarize their ideas about the reading.
4. **Next, their partner should respond** (agreement, disagreement, clarification, offering a suggestion, etc.).
5. **Then the students switch** and the other student shares his or her significant comment.
6. Next their partner has the opportunity to respond.
7. **Students read several short paragraphs,** alternating back and forth about their significant summarizing statement and they then listen to one response.
8. To add accountability to the conversations, **you may want to ask students to write out their summarizing statements** before reading them out loud. This way you can track the quality of their ideas to make sure students are getting the gist of the reading.

K and ELL Scaffolding:

K and ELL students will need plenty of support as they practice structured discussions with their peers. Have these students work on only one sentence frame like "expressing an opinion" four to six times before trying the next sentence frame.

Why the Structured Discussions Intervention Works

Structured Discussion works because many students have few opportunities to engage in formal conversations. The sentence frame gets students started on the right track. Structured discussions also cause the students who are listening to pay more attention, because they have a specific responsibility that they are required to contribute to the discussion.

Structured Discussions Progress Monitoring

Structured Discussions are easy to monitor. The teacher should circulate around the classroom and listen to students to see if they are using the sentence frames as designed. Once students have started to use the Structured Discussion sentence frames effectively, they should transition to engaging in discussions without using the sentence frames. It is extremely impressive to see a class of students discuss things positively and appropriately.

- See how quickly students can transition from a sentence frame to engaging in a quality discussion without the sentence frame.
- Identify how long students can stay in a structured discussion before they want to break down and get off topic.
- Ask students to practice at home if they need to make more progress.

It is sometimes necessary to make sure that students' understand the concepts being taught. A quick but very effective means is to ask students to use their white boards. Call out a verbal response like, "No thank you." Ask the students to identify what kind of a response it is by writing the title of the response on their whiteboard: "disagreement." Reverse the practice by writing a component of discourse on the white board and then ask the children to write what it would look like. All progress monitoring should result in the students' determining how well they or how much they need to work on this aspect of language. You can use a stoplight as a metaphor—green for "I get it," whereas yellow would represent, "I think I need more help." Red stands for, "I don't understand." You could graph the number correct answers out of the number of questions asked, "I got 7 out of 10 correct." Graph is by row, individually, or in teams.

Speaking Intervention #4: Fluency Phrasing

Students seem to be okay with the rate and accuracy portion of fluency. They often leave out the very important third part of fluency, which is prosody (Figure 5.10). Pinnell and Scharer (2003, p. 5) write, "If we value phrased, fluent reading, then we need to teach for it. Children will not automatically develop a sense of phrasing and fluency."

Figure 5.10 Three Parts of Fluency

- *Rate:* The number of words read in one minute for a given passage
- *Accuracy:* Reading without "missteps"; reading words, punctuation, and expression congruently.
- *Prosody:* This phenomenon is characterized by inflection, tone, and expression that is clearly articulated by the way the author wrote the words. It is the theater of being able to read well.

Fluency phrasing emphasizes all three aspects of fluency, yet it places the most importance on prosody. Prosody is a word that may be new for many teachers. We find that only a quarter of K-2 teachers at our workshops know what this word means at a four or five level of understanding (Figure 5.11).

Figure 5.11 Three Parts of Prosody

- *Inflection:* This is how we accentuate a word because of the meaning it has in the context of the print.
- *Tone:* This relies on the ability to use your voice in various ways—soft when speaking to a baby, louder when at a baseball game or if a car nearly hits the stroller. Tone reflects the readers understanding of how the words tell the story.
- *Expression:* This is the theater of reading aloud. By using all of the tools in punctuation, context clues, and knowing vocabulary, the reader incorporates all that is known into a presentation that should reflect what the author intended.

What the Fluency Phrasing Intervention Looks Like

Students need plenty of practice reading text. Effective instruction emphasizes prosody that is the inflection, expression, and intonation used to speak prose out loud.

- Raise voice at a (?) question mark.
- Lower voice and pause at a (.) period.
- Shorter pause at a (,) comma.
- Voice change at (" ") quotation marks.
- When reading out loud for fluency, speak smoothly if there is no punctuation.
- Read emotion words with congruent inflection and facial expressions.

How the Fluency Phrasing Intervention Works

1. **Put students in pairs** with a partner of similar ability.
2. **Provide students reading passages** that contains a variety of sentence punctuation endings (?, ., !).
3. **Ask students to read sentences out loud** in partner reading and practice accentuating the proper phrasing.
4. **Give students a checklist of sounds with illustrations** so that they might see their progress.
5. **After practicing this strategy with a partner**, students can use a fluency phone to get additional practice.
6. **Students can then transfer their abilities** by speaking a dialogue between students, reading poetry, or presenting a play in class.

> **K and ELL Scaffolding:**
>
> Model for the students in a whole class or small group examples of fluency phrasing through guided reading and then ask the students to repeat back out loud the same phrase with the same emphasis, inflection, and tone that you used.

Why the Fluency Phrasing Intervention Works

This strategy works because a primary connection between decoding and comprehension is effective development of fluency. Fluency is much more than just rate and accuracy. The key component that reveals a student's progress toward comprehension is their ability to use proper phrasing while reading at an appropriate rate and with accuracy in a way that develops automaticity. Another word for proper phrasing is prosody. As students listen to proper phrasing they will be able to speak with appropriate phrasing and demonstrate their comprehension.

Fluency Phrasing Progress Monitoring

Have students listen to each other in pairs, and circulate throughout the room to listen to their progress.

- Listen to see if students pause effectively at commas and end sentences with the right inflection.
- Provide students opportunities to read poetry (their own or others') in class.
- Ask students to read dialogue or do short plays with dialogue that has plenty of emotion.

Speaking Intervention #5: Academic Talk

Once students have the basic conversational skills involved in structured discussion discussed previously, they need to develop their ability to effectively engage in academic talk. Wolfram, Adger, and Temple (1999) note that academic talk helps all students: "All children need linguistically rich classrooms in all subject areas to develop expertise in literacy and academic talk."

Conversations occur with friends, in the home, and around the neighborhood. Yet the instructional conversations that comprise academic talk are very important for students to develop.

- Tapping Prior Knowledge
- Predicting
- Picturing
- Internal Questioning

- Making Connections
- Generalizing
- Forming Interpretations
- Clarifying Issues
- Relating our Learning
- Summarizing
- Reflecting on our Learning

What Academic Talk Intervention Looks Like

Academic talk is content-area talk that is scaffolded by the classroom teacher to support academic processes and get positive academic results. Geluykens and Kraft (2008) emphasize that a significant gap that affects many students focuses in on academic talk: "The first gap concerns the neglect of academic talk."

Consider the following chart from Girard (2005) (Figure 5.12):

Figure 5.12 Percent of Time ELL Students Speak English in Class

4% of time is spent speaking casual language 2% of time is spent speaking academic language

Students are immersed in language all day long (Figure 5.13 on page 116).

How the Academic Talk Intervention Works

Classroom discussions between peers in small groups or pairs provide students with important opportunities to develop their ability to learn. When students produce language and communicate with their peers, the academic talk helps students become more consciously aware of the important cognitive strategies that develop academic literacy. Like the speaking habits that we want students to develop, cognitive thinking habits need to be developed by students as they engage in academic talk.

1. Students are placed in groups of four.
2. **Each student is provided and carefully reads a passage** or piece of a content-area passage.
3. **Students highlight the main idea** or gist of the topic.

Figure 5.13 **Academic Talk Frames**

Tapping Prior Knowledge ♦ I already know that… ♦ This reminds me of… **Predicting** ♦ I guess that… ♦ I believe that… **Picturing** ♦ I can see… ♦ I can imagine… **Clarifying Issues** ♦ I have a question about… ♦ I'm unsure about…	**Making Connections** ♦ This is like… ♦ This reminds me of… **Asking Questions of Others** ♦ I wonder why… ♦ What if… **Forming Interpretations:** ♦ What this means to me is… ♦ The idea I'm getting is… **Reflecting on learning** ♦ The most important idea I learned today was… ♦ The most challenging part of our learning was…

4. **Students use academic talk sentence frames** to elaborate on academic talk (i.e., making connections, generalizing, clarifying issues).
5. **Students use academic language** and each take turns sharing their sentence frame responses about their reading passage with their peers.
6. **Students summarize together** and create a shared meaning that they generated from sharing and connecting the jigsaw pieces.

> K and ELL Scaffolding:
>
> This strategy will be difficult for kindergartners to execute effectively. They should master Speaking Intervention #3: Structured Discussion before students should attempt Academic Talk. This intervention is better suited for first- and second-grade students.

Why the Academic Talk Intervention Works

Academic talk is intended to extend, clarify, state exceptions, give examples, make connections, amplify understanding and more within targeted content area discussions.

Progress Monitoring Academic Talk

- Listen to students as they engage in academic talk in different subjects to make sure they are effectively understanding the content.
- Check to make sure that students use the sentence frames and ask students to use more than one academic talk topic to increase rigor.
- Once students practice the strategy numerous times, take away the academic talk prompts and see if they can use academic talk spontaneously.

Summing It Up

Students need more opportunities to express their thinking and ideas through classroom conversations. The Sound Muncher intervention is highly engaging and helps students maintain their motivation as they practice their phonemic sounds and phonics blends. In fact, the intervention can work for a variety of basic learning strategies (Word Cards, etc.). Fluency phones are an excellent way to get the entire class to actively speak fluently to improve their rate and accuracy. Fluency Phrasing helps students work on prosody or the proper expression and intonation used when speaking. Structured discussions help students practice basic conversation patterns that will benefit the students in the class, on the playground, and in the future. Academic Talk helps students see that certain conversational processes support learning and meta-cognitive understanding. Students who can speak effectively about academic topics will be better prepared to write about academic topics.

Reflection

1. How often do your students receive opportunities to engage in structured conversations about school related topics?
2. Are all of your students reading fluently at grade level?

6

K-2 Intervention Writing Strategies

"I love writing. I love the swirl and swing of words as they tangle with human emotions."

—James Michener

Mark dreads writing assignments that are longer than a short sentence or two. He has lots of ideas in his head, yet it seems like they never make it down on paper the way he intends. His handwriting is sloppy and it always feels like a chore to take the time to write. He likes to talk in class and is a frequent participant in class discussions. His classmates view Mark as a leader. He always seems to be picking the teams at recess time, yet in class he avoids writing assignments. Like many of his peers, Mark is a reluctant writer who may someday become a resistant writer.

Writing Routines

It is important for students at an early age to find ways to express their thinking through writing. Writing routines that emphasize consistent interventions should be a part of students' learning activities. Stronge, Tucker, and Hindman (2004, p. 97) note the importance of developing instructional routines in writing:

> The use of instructional routines shifts the emphasis from the how the instruction is delivered to the content that is addressed.

We define writing as their personal effort with words that produces original text. After we are sure the students have grappled with what writing is, then it is necessary to help them define who they are as authors. We have found that this is one of the most important aspects of the writing process. Letting every student know that they have the capacity to be authors requires that you define what a writer does to become an "author." We let students know that a writer's job is to put together their own thoughts, organize them, and put them into words using their creativity, language conventions, and writing mechanics. To be a writer, students have to be thinkers, organizers, and capable of following structural elements. When students agree that they can do these things, we introduce the concept of creativity. Writing is the ultimate creative act. The discovery of how a student can be creative with their writing is what makes all of this so worthwhile. It is also a great mirror for future careers and expertise.

When we start the process of teaching writing with students and ask them how it might be defined, they invariably define writing in terms of the subskills that are required to do it well. They will call out that "writing is sentences with periods," or "writing is about putting words on paper." Some will say that "writing is about knowing rules." More sophisticated students will guess that writing is linked to a "purpose." For example, writing "tells a story" or it "describes something," perhaps it someone's "opinion" or even "educates" you. With this perspective, the students begin to tap into what good writing will do for a reader. They have started to define four of the basic benefits of reading and the effects of great writing. Most readers enjoy the following:

1. A good story
2. Vivid descriptions that make pictures in their minds
3. Connecting specific ideas from what they read
4. Finding out what other people think and feel

Troia, Shankland, and Heintz (2010, p. 241) add:

> It is, therefore, important to design interventions that may be used in teachers' classrooms, integrating new writing interventions into their existing practices and aligning them with curriculum content standards and materials.

Gersten and Baker (2001, p. 254) cite several intervention areas that will improve student writing:

- Explicit teacher modeling of the writing process and composing strategies
- Peer collaboration and teacher conferring to gain informative feedback
- Use of procedural prompts (e.g., graphic organizers, mnemonics, outlines, checklists) to facilitate planning and revising
- Help students who have poor hand writing (e.g., dictating)
- Self-regulation (e.g., self-statements and questions)

As our students make the connections between reading narrative and informative text with their own writing, comprehension will improve.

Writing Intervention #1: Spelling and Writing

As early as the first day of kindergarten, students can engage in spelling as a way to develop their writing skills. Asking students to write words before they know how to spell the words perfectly helps them understand the importance of the writing process. K-2 students can expand their vocabulary by spelling and writing new academic words that they hear at school. The more students write, the more words they will be able to add to their personal dictionary. The strategy of invented spelling allows young students to engage in the writing process by trying and exploring ways to express their ideas with words. Invented spelling gives students the opportunity to write letters, words, and phrases without feeling that they have to be a perfect writer the first time. Invented spelling gets students engaged in the process of writing and it supports the development of fundamental skills related to reading, writing, and learning. The Learning First Alliance (1998, p. 14) noted:

> Creative and expository writing instruction should begin in kindergarten and continue during first grade and beyond... Research shows invented spelling to be a powerful means of leading students to internalize phonemic awareness and the alphabetic principle...

It should also be noted that even though young students benefit from the freedom to use invented spelling to spell and write words to the best of their ability, the students also need eventually to learn to spell the words in the vocabulary accurately. Just as older students benefit from writing a rough draft that is filled with many mistakes and then going through the writing process to edit, polish, and improve their writing, younger students benefit from invented spelling before going on to correct spelling.

What Spelling and Writing Looks Like

The invented spelling strategy is different than a typical Friday spelling test. This strategy looks into the process and progress of student writing as it relates to sound-letter relationships, and it provides a program for evaluating student writing in the earliest stages. Snow, Burns, and Griffin (1998, pp. 323-324) state:

> [Spelling] instruction should be designed with the understanding that the use of invented spelling is not in conflict with teaching correct spelling. Beginning writing with invented spelling can be helpful for developing understanding of phoneme identity, phoneme segmentation, and sound-spelling relationships. Conventionally correct spelling should be developed through focused instruction and practice. Primary-grade children should be expected to spell previously studied words and spelling patterns correctly in their final writing products. Writing should take place on a daily basis to encourage children to become more comfortable and familiar with it.

Invented spelling as a strategy should be used to engage students in writing and eventually lead to correct spelling as students get comfortable with gripping the pencil, writing their letters, sounding out words, and understanding phoneme-letter relationships. The greatest benefit of invented spelling intervention may be its ability to help us assess students' levels of writing and reading. Consider the following method for evaluating invented spelling or student writing at the early stages (adapted from Gentry 2006, p. 192) (Figure 6.1).

How Spelling and Writing Works

When students begin the earliest stages of the writing process with invented spelling, it allows them to express themselves creatively without focusing on being perfect. At the same time students need to learn how to spell words correctly. The following steps can help students write words through invented spelling while also helping them write words correctly.

1. Let students know that you want them to **do their best to spell** some words that may be new to them.
2. **Ask students to get out their white boards** with their white board pens and erasers.
3. Make sure **students are in the ready position** to write on their boards.

Figure 6.1 **Early Stages of Student Writing**

Level 0: No Letter Use
Student draws and scribbles while pretending to write

Phase 1: Non-Alphabetic Writing
Student writes in random letters and attempts to label things

Phase 2: Partial Alphabetic Writing
Student writes simple stories (a couple of lines); attempts several genres (lists, commands, messages, directions, labels, writing on maps)

Phase 3: Full Alphabetic Writing
Student writes simple stories with a beginning, middle, and end

Phase 4: Writing in Chunks of Spelling Patterns
Student writes more elaborate stories first-then-next-last, writes in varying genres with academic language

4. **Say the first spelling word, and ask the students to say the word out loud together** as a group. (With younger students or ELL students you may want to repeat this process several times to make sure that all of the students are saying the word correctly.
5. **Say the word in a sentence** so that students have an example and a context for the word.
6. **Ask students to write the word on their white boards** the best they can and to sound out the letters or syllables as they write it.
7. After 20 seconds or so, **ask students to hold up their white boards** and show you their spelling.
8. **Review the spelling on the white boards** to get a clear idea of which phases the students are operating in (see Figure 6.1) and make mental notes of which students will need additional help during small-group time.
9. **Hold up your teacher white board with the correct spelling** on it.
10. **Ask students to write the word with correct spelling** below their first attempt at the word.
11. **Go to the next word and repeat the process** by saying the word, asking students to say it together as a group, and then asking them write it on their white boards.
12. **Give the students approximately 10 to 12 words to use invented spelling** and then spell correctly on their white boards.

K and ELL Scaffolding:

Kindergarten and ELL students can begin with invented spelling by writing simple words like (do, act, go, read, add, and save).

For older students in first or second grade, the following word lists or writing activities can be used for daily writing. Again if students misspell some of the words in the process it is okay, because you can always go back and edit and correct misspelled words.

The following list provides ways to get young students to write:

1. Lists (about groceries, items in the class, favorite foods, etc.)
2. Invitations (for birthdays, special events, parties, etc.)
3. Notes Home (about learning, school activities, etc.)
4. Letters (to the President, principal, friends, etc.)
5. News Reports (about the many "happenings" in the classroom, on the playground, or at home, etc.)

In this way, writing can be accomplished three or four times a day.

Why Spelling and Writing Works

The Spelling and Writing Intervention works on many levels. Students need lots of initial practice learning to write letters, to spell out sounds, and to construct words. Invented spelling gets students engaged in the writing process without worrying about being perfect writers. This strategy helps students get plenty of practice writing new vocabulary words. Transferring the sounds in their head into writing on their paper helps students develop phonemic awareness and sound-letter relationships. Plenty of daily practice writing helps develop the alphabetic principle for students.

Spelling and Writing Progress Monitoring

The following writing phases provide a clear idea of what level students should achieve to write at grade level (Figure 6.2).

Students who struggle to reach the phases of writing that match their grade level should receive small-group support.

- Model for the student key elements of the writing process.
- Give students plenty of practice writing lists or other related writing tasks.
- Show students examples of previous students' work that models good writing.

Figure 6.2 **Assessing Student Writing Levels**

Level 0: No Letter Use Recognize that this student will need interventions to help them progress
Phase 1: Non-Alphabetic Writing Provide intervention if not observed by mid-kindergarten
Phase 2: Partial Alphabetic Writing Provide intervention if not observed by end of kindergarten
Phase 3: Full Alphabetic Writing Provide intervention if not observed by the middle of first grade
Phase 4: Writing in Chunks of Spelling Patterns Provide intervention if yet to be observed by the end of first grade

Writing Intervention #2: Shared Writing

Shared writing is extremely helpful to young students in the primary grades who are still developing their own concepts of writing. Shared writing can be extremely revealing to teachers as they see the types of challenges and questions students may have as they proceed through the writing process in a specific content area. Students learn to write well when they see the connections between what they say, hear, and see. With young children, shared and interactive writing are ways to bridge what students say, hear, and see with how it looks in print. Pressley and Billman (2007, p. 135) add, "Writing strategy interventions were found to yield rather large gains in writing performance…"

Shared and interactive writing provides the opportunity to practice sight words and see sentence structure. It models the connection between saying and seeing the language in print. Making writing connections from reading experiences is like writing a good newspaper article. Routman (1994) lists several benefits of utilizing the shared writing strategy with students. Some of these include the recognition that shared writing:

- Reinforces and supports reading as well as writing
- Makes it possible for all students to participate
- Encourages close examination of texts, words, and options of authors
- Demonstrates the conventions of writing-spelling, punctuation, and grammar
- Focuses on composing and leaves transcribing to the teacher

Students appreciate that the writing process can be a shared process rather than a solitary effort.

What Shared Writing Looks Like

Let's take a look at shared writing that is designed to help students move toward greater independence.

Henn-Reinke and Chesner, (2006, p. 112) write, "In a shared writing experience, as teachers model correct usage and writing mechanics, it is beneficial if they do a think-aloud, describing why they are writing, spelling, and formatting in a particular manner."

Students may soon recognize that through consideration and discussion their collective paper can be clarified and revised to improve the meaning of the message. Language is a shared experience and sharing the construction of writing can be powerful as they learn to develop their ideas and write collaboratively.

How Shared Writing Intervention Works

Students enjoy working together in the creative process of writing. With help and support, students will begin to see the important components and processes for writing effectively.

1. Under your guidance, students should work together to **select a topic.**
2. Everyone works together with teacher assistance to **generate a thesis statement.**
3. Students should work together to **choose and then write three topic sentences.**
4. After the main ideas and core content are determined, students should work together to **write an attention grabber.**
5. You can act as scribe and **write down transitions sentences between paragraphs.**
6. Students should work together and **decide on at least three supporting details for each paragraph's topic.**
7. Work with students to **summarize the significant points contained in the writing.**
8. With your support, **help students edit, revise, and finish their writing.**
9. **Ask students to give their own evaluation** and assessment of their collective work.
10. Finally, invite students to provide a **formal presentation of their collective writing** to the librarian, principal, or parents.

While the steps above are for a shared informative writing activity, shared writing can also be used for narrative writing.

During shared writing it is important to model:

1. Begin by **establishing the purpose** for the writing and discuss how this will determine the topic, organization, and main points.
2. Next, focus your instruction on one or **two specific writing principles** or processes that you want your students to learn.
3. **Explain out loud the decisions and choices** you make as you write.
4. Rehearse the sentences before you will write down by **saying the sentences out loud**.
5. Encourage and **model the habit of using proper punctuation** or go back and fix punctuation as your write.
6. Constantly **reread your writing** to help your students determine the flow from one sentence to another.
7. Stop from time to time and **consider out loud other academic language** or words which may more precisely convey your message.
8. Remember to **move at a thorough yet appropriate pace** to ensure you keep students' attention.
9. **Encourage suggestions from students**, checking for misconceptions and providing further explanation.
10. **Rewrite certain phrasing** from time to time to focus students' attention on selecting specific word choice.

K and ELL Scaffolding:

Younger, new students definitely will need you, the teacher, to act as scribe for the story. For younger students writing a simple narrative story with a beginning, middle, and end will feel like a significant accomplishment.

Why Shared Writing Works

The text that is generated from collaborative writing becomes a recorded document of class activities and learning. This permanent record can be shown to other classes or groups and serves as a student model of writing within the content area. Through collaborative writing, students contribute their own ideas to the content of the text, and they can see that writing is an interactive process of editing, revising, and refining ideas. In time, each individual student will begin to develop their own understanding and confidence in using the process for the various content area writing assignments.

Program Monitoring Shared Writing

Just as students love having the teacher read to them, K-2 students also appreciate cocreating writing with their teacher. As you guide and lead the students in the important components of writing, the conversation will help you discover how well students are doing as they articulate their ideas and thoughts about writing.

- You can place students in small groups of four to see if the students in these groups are able to generate their ideas for topic sentences, thesis statement, and attention grabber.
- For students who appear to be contributing little to the whole class process of shared writing, pull a small group together and work specifically with them to make sure they are understanding the key components of the writing process.
- Ask students to work together with a partner to do their own paired shared writing to see how well they can do.

Make sure to have fun in this process, so that students learn from the earliest years that writing can be an enjoyable process.

Writing Intervention #3: Make 'n Break Writing

Writing for students in the earliest grades of school can be a challenging task. Just getting students to write legibly and hold their pencil correctly can be a challenge. At the same time, students benefit from constructing their ideas and thoughts into writing. As students are able to express themselves effectively in writing, their confidence as learners will significantly increase. The challenge is how to scaffold writing in a way that students can engage in the basic components of writing even though they have limited skills. Analyzing the way that language is constructed to convey meaning can help students analyze the words and negotiate meaning.

What the Make 'n Break Writing Intervention Looks Like

This intervention looks a lot like a puzzle that needs to be put in the right order. Students benefit by considering the component parts that make up written communication. As students analyze the words that make up sentences, they can put word cards or sentence strips together.

- **Word Strips** (Words written on strips of paper that can be moved around and manipulated by students.)
- **Sentence Strips** (Sentences written on strips of paper that can be moved around and manipulated by students.)

Students look to bring a cohesive understanding to a jumble of words by organizing them into sentences and paragraphs in their proper order. This strategy works because students already have all of the components of the message to begin the process, so this intervention is effectively scaffolded for young learners. The sentences can be extremely basic with four-word sentences for new kindergartners, like "Nate ate the cake."

Make 'n Break Word Cards

Provide the following word strips to students and ask them to write them down in the proper order.

ate	*Nate*	*cake.*	*the*

Students should be able to notice that the capitalized word goes first and the word with a period after it goes last. This simple sentence should be correctly reordered and written as:

Nate ate the cake.

Providing students with the correct words already written out in word strips allows young students to easily manipulate the words to develop complete sentences. The Make 'n Break strategy can also help students write brief paragraphs (Figure 6.3).

The sentences should be reordered to make a complete story.

1. Once upon a time there was a princess.
2. She kissed a frog.
3. The frog turned into a princess.

Figure 6.3 **Make 'n Break Sentence Strips**

She kissed a frog.
Once upon a time there was a princess.
The frog turned into a prince.

The teacher can explore with different sentences to see students' proficiency with constructing a variety of sentences. Use sentences that are questions, end in exclamations, have a comma in the middle, or make up a declarative or imperative sentence ending in a period.

How the Make 'n Break Writing Intervention Works

1. **Write out a sentence on the board** and read the sentence with the students.
2. **Place word cards on the board** to show the individual parts of the words that make up the sentence.
3. **Help students recognize the parts of the sentence** by talking to students about the "who" or "what" that is the subject and the "action" or what "they did" in the sentence.
4. **Discuss concepts with students like the first letter** in the first word is capitalized, the last word has punctuation after it, and other writing conventions.
5. Next the strategy really begins, **as you ask students to team up with a neighbor** for a partner activity.
6. **Then take another sentence** where all of the word cards are mixed up and out of order and ask students to discuss with their partner the proper order to make a complete sentence.
7. **After given time to work out the word order,** a pair of students should come to the board and arrange the word cards in their proper order.
8. **Students should explain out loud the process** that they went through to construct meaning from the jumbled words and justify their decisions.

K and ELL Scaffolding:

Young students should be provided simple word strips of familiar words that they can reorder into correct sentences. Provide as much help as students need to be successful.

Why the Make 'n Break Writing Intervention Works

The kinesthetic nature of tangibly touching the word cards or sentence strips helps students to connect with the writing purposes. Because students can work with a partner, they can have support and help in the process of comprehending meaning. English Language Learners often construct sentences differently than we do in the United States (for example, the adjective may come after the noun in other countries), so the Make 'n Break strategy can be extremely beneficial to teach sentence structure. Because students can look at the word strips or sentence strips and discuss their thinking verbally, it is extremely helpful to them to think and meta-cognitively rehearse their understanding out loud.

Make 'n Break Writing Progress Monitoring

The Make 'n Break Writing intervention is easy to monitor student progress, because you can actually watch the students manipulate the order of the word cards or sentence strips and hear them negotiate with their neighbor how they believe the sentences or paragraphs should be constructed.

- Listen to students as they discuss their decision-making process for determining the order of the word cards or sentence strips.
- Focus on one skill at a time and emphasize with the students the benefit of learning that skill (i.e., capitalization goes at the beginning of the sentence, punctuation goes at the end of the sentence, the difference between facts and opinion, or the difference between a main idea and a supporting detail).
- Give students quick feedback as to the success of their answers. This way the students will recognize when they are on track or they can ask questions for further understanding.

Organizing and constructing the sentences helps students remember information they are learning.

Writing Intervention #4: Picture Word Inductive

The Picture Word Inductive strategy is a powerful strategy that works well for young students, particularly English Language Learners. The primary purpose of the strategy is to develop students' word knowledge, sentence construction, and paragraph structure. Through the use of pictures and posters, students work inductively to learn vocabulary words. The model was developed by Calhoun (1999, p. 21):

> The Picture Word Inductive Model is an inquiry-oriented language arts strategy that uses pictures containing familiar objects and actions to elicit words from children's listening and speaking vocabularies.

These vocabulary words serve as the basis for developing background knowledge about a topic that is new to the students. This strategy works extremely well for students who are new to our language and customs. It also works extremely well for students who are new to school in the primary grades. The vocabulary word lists that students generate should be combined with other phrases to create sentences and paragraphs. Like the other strategies in this book, students are the central focus of this strategy and they learn to direct their comprehension and learning. It is designed to support students as they become detectives and discover how to "crack the code" of language learning.

What the Picture Word Inductive Intervention Looks Like

They say that a picture paints a thousand words. The Picture Word Inductive intervention is an inquiry-based strategy where students provide information, ask questions, and build background knowledge as they analyze a picture containing familiar objects and actions. This strategy was especially designed for K-2 classrooms and the challenges that young students bring to school. The Picture Word Inductive strategy helps students inquire about words, discover language structure, develop their vocabularies, and increase writing abilities. This strategy works extremely well with English Language Learners or any other struggling student who has limited academic vocabularies. The model is designed for children to learn inductively. They group information together and begin to recognize patterns within the information. This generalization of understanding can help provide a foundation for phonetic analysis and writing structure.

The primary objectives of the Picture Word Inductive intervention are that students can construct writing from the words they identify from a picture. Students should:

1. Identify objects and actions in pictures with precise vocabulary while building background knowledge.
2. Categorize newfound words into generalized patterns.
3. Finally, they can use their expanded vocabularies and understanding to write about their newfound understanding.

Figure 6.4 **Picture Word Inductive Sample**

What does it look like? Students generate from pictures the words and sentences that make up a picture and studying those words and sentences. Lessons can last as short as 15 minutes for some students or take up an entire one-hour reading block for a class of ELL students. Although the strategy can work with a singular item in the picture, it works best with a complex picture with a variety of items in the picture with different things happening. A collaborative strategy where students collectively construct their own understanding from a picture is best (Figure 6.4).

Words that students generate from the picture:

- Horse
- Sheep
- Cow
- Barn
- Fence
- Grass
- House
- Goose
- Chicken
- Rooster
- Dirt
- Pig
- Tree
- Window
- Chicks
- Roof

From the word lists the students generate, they can then categorize the words into different groups. Students benefit from finding word patterns, because it helps them organize and remember the information. The examples in the next section come from a variety of pictures (Figure 6.5 on page 134).

Figure 6.5 **Student-Generated Categories**

Student-identified word categories	Students explain their word categories
reindeer, road, racetrack	♦ All begin with *r*
book, books, black, board, blue	♦ All begin with the letter *b* and have one part
picture, people, person	♦ All have *p's* ♦ All have *p's* at the beginning ♦ All have *p's* at the first letter and two parts ♦ All have *p's* at the beginning and they all have *es*
girl, boy, child, children, people, person	♦ All humans ♦ All are names for people when we don't know their names ♦ All are people
black book, yellow book, blue book	♦ All have *book* ♦ All have the color of the book ♦ All have two *o's*
green hat, black book, red T-shirt, yellow book, blue book, white board, gold letters	♦ All have color words
necktie, white board, child, children, shiny book, hair, red T-shirt, girls, pictures	♦ All have *i's*
boy, girl, child, children	♦ All are names for kids

Students can take basic words and use the following intervention approaches to come up with original writing. Wilhelm (2002) uses these intervention perspectives to extend thinking:

- ♦ New Circumstances
- ♦ New Context
- ♦ New Age or Time Period
- ♦ New Narrator or Genre
- ♦ New Conclusion

So students can take a list of words and use these intervention perspectives to create new and original ideas while they are using common vocabulary words.

How the Picture Word Inductive Intervention Works

This strategy may take several days to go through all of the steps of the process with young students. The pictures selected can be about science (a farm), social studies (capitol building), mathematics (a building with measurements), and language arts (a fairy tale).

1. **Find a picture.** Make sure that the picture has lots of objects, actions, and dynamics happening in the picture.
2. **Place the picture on a piece of butcher paper.** There should be space enough to draw arrows to identify different objects, actions, and dynamics in the picture.
3. **Invite students to identify things they see in the picture.** Students will most often identify objects first, so prompt them to also look at actions and relationships in the picture
4. **Label the picture parts identified**. Write the word and then draw an arrow or line from the identified object to the word, say the word out loud, and then ask the students to spell the word and say the word out loud together.
5. **Read and review the picture word chart aloud.** After identifying many objects and actions, go over all of the words and discuss relationships between things.
6. **Invite the students to classify the words.** Have students put the words identified into categories (i.e., kindergarten students may use beginning consonants, rhyming words) to emphasize with the whole class.
7. **Review the picture word chart.** Have students say the word, spell it, and then say it again.
8. **Ask students to come up with a title for the picture word chart.** Ask students to think about the information on the chart and what they want to say about it.
9. **Add more words to the chart.** Ask students to take another look at the chart and see if they can add more words.
10. **Model putting the words into good sentences or a paragraph.** As you construct sentences, talk about capitalization, punctuation, word choice, etc.

11. **Place students in teams.** Students can be partnered or in groups of three or more to share understandings and negotiate meaning.
12. **Ask students to generate a sentence, sentences, or a paragraph.** Kindergarten students may feel good to construct one or two simple sentences with help. Second graders can be asked to construct short paragraphs that summarize the objects and events happening in the picture.
13. **Have students read the sentences and paragraphs of their peers.** Students should look for patterns and offer suggestions to extend and elaborate others' ideas.

> **K and ELL Scaffolding:**
>
> With new students, start out with pictures that will be easy to identify. Encourage them to find very simple patterns like words that start with the same letter.

Why the Picture Word Inductive Intervention Works

It is important for students to engage and play with words that make meaning for them. This strategy allows students to first identify words and then to play around with the words and organize them before writing and constructing meaning out of the newly acquired words. Students see a tangible picture and visually connect it to a new word in their vocabulary. Students hear and see the words spoken and written repeatedly and correctly. Teachers can emphasize any sound or symbol relationship that they want to work on at that time. Students see the words and connect the words within a context created by the picture. Students classify and organize information within their mind in ways that make sense to them. They get to hear the variety of ways their peers classified and organized the information. When writing the sentences, students learn to consolidate and elaborate on information.

From this picture the children learn to associate pictures with words.

- They learn that the picture can be used as an illustrated dictionary or reference.
- They learn to classify words (looking for patterns in words, decoding strategies, similarity in endings). We can ask students to reclassify words several times (developing their skills of inquiry, and then develop additional teaching lessons).

- They learn to classify sentences, as well as to distinguish two syllable words and use lessons to add adjectives (descriptors to words. . . eg., color).
- They learn to create titles, topic sentences, and paragraphs. We can ask questions of the students to facilitate deeper thinking.

This program allows students to be successful quickly, the more words a child adds to their sight vocabulary, the quicker they can start making connections that will lead to reading. When a student has developed a vocabulary of about 450 words, many picture storybooks are available to them. At 50 sight words, phonics skills are facilitated. This strategy does an excellent job of introducing new vocabulary and concepts to students through a process of discovery. The strategy actually integrates speaking, reading, and writing skills together. The strategy is student centered, yet led by the teacher to make transitions and help students make connections between content and comprehension. The strategy helps students make a speaking-reading-writing connection.

Progress Monitoring for Picture Word Inductive Intervention

Because the teacher is involved with students from the beginning of this strategy until they complete their writing, it is easy to see how students are doing with the process of identifying, categorizing, and finally writing.

- Throughout the identifying process, listen to the words that students are noticing to see what cognitive connections they are making. The more elaborate actions and relationships the students notice, the better.
- Observing students as they engage collaboratively in the classification process provides a terrific opportunity to see how students organize their learning and negotiate meaning with others.
- Monitoring student writing provides the best evidence of students conveying the meaning and learning what is going on in the students' minds.

The Picture Word Inductive intervention will help students generate writing as they construct from a picture the words and eventually the ideas and sentences that develop authentic writing.

Writing Intervention #5: Writing Summaries

Summarizing requires students to put in their own words a shortened version of information. Summarizing is the process of reading selections of text and then putting the central message into your own words. Summarizing reduces the content of the text into the key elements that convey the central meaning of the text. Summarizing is a process of analyzing and reviewing information. A summary distills the gist of the information being communicated by the author to the reader. The reader should sift through the information to grasp and comprehend the main idea or points being made by the writer. In a day when written text in its many forms is becoming more prevalent, students' ability to summarize is an essential skill. Wormeli (2004, p. 2) points out, "Summarization is one of the most underused teaching techniques we have today, yet research has shown that it yields some of the greatest leaps in comprehension and long-term retention of information."

Although it is easier for students to summarize written text because they can go back and analyze the text, the process of summarizing helps students comprehend spoken communication in addition to written communication. Summarizing should help students see the main structures of knowledge that support comprehension and learning where students recognize the parts and the integrated whole of meaning (Brophy & Good, 1986).

What Writing Summaries Looks Like

A summary is a consolidated version of a text. It contains the main points in the text and is written in each students' own words. It is a mixture of reducing a long text to a short text and selecting relevant information. A good summary shows that you have understood the text. Walker and Rivers (2003) note the benefits of writing summaries from text for students: "Writing summaries of texts they have read helps them learn to use writing to elaborate on what they are reading."

Summaries of written text are valuable indicators that the reader has understood the content conveyed in the text. Zamuel and Spack (1988) point out that students should learn, "the complex activity to write from other texts," and summarizing should be "a major part of their academic experience." Students should be given ample opportunities to read text and summarize the significant points made by the author. Summaries can synthesize a variety of different texts, yet reading rich, vibrant text makes for more exciting and interesting summaries. Students need to clarify the main ideas conveyed in the text. Summaries should organize the main ideas in a similar order that the author structured the information. As students learn to summarize, they will strip away the extraneous information and extra verbiage, while they focus in on the heart of the matter. Through summarizing students strive to capture the main ideas and critical supporting details that are at the center of the essential meaning being communicated. Students who can summarize effectively demonstrate their ability to comprehend information in powerful ways.

- Focus on the main idea or ideas
- Identify the key supporting details
- Comprehend the overall meaning
- Write down only enough to demonstrate understanding of the essential meaning

Summarizing is a very challenging process to teach well. Students need repeated opportunities to engage in the strategy and show their ability to summarize different types of text. While we ask students to summarize in its various forms frequently, as teachers we seldom explicitly instruct students in the process of summarizing. Throughout their educational careers, students will be asked repeatedly to summarize in its various forms to demonstrate comprehension and identify the essential meaning of written and verbal communication. Although this intervention takes time and patience to develop in all students, the benefits are significant and long lasting. The steps to success listed below will be well worth the effort needed for students to be successful. Copies of simple-to-read newspaper articles make excellent texts for young students to analyze and summarize. Students will enjoy the idea that they are able to get the gist of information from a newspaper that may seem like a large and looming challenge to a 5-year-old to a 7-year-old.

How Writing Summaries Works

Select a text to be summarized (to begin with, choose a text or subject with which they are already familiar) and instruct the students as follows. It may be helpful to list these instructions on the board or flip chart so students can refer to them as they work.

1. **Read the text** through for general understanding.
2. Read the text a **second time,** analyzing key parts.
3. **Underline** key words, phrases, and vocabulary.
4. **Circle** the main idea(s) (usually the ***who, what,*** or ***why*** questions).

5. Put a **box** around supporting details that give more information about the main idea.
6. Use underlined, circled, and boxed words to **write a summary with only 15 words**.

Younger students will need help writing key phrases and sentences to create the summary. Summarizing can challenge students to write using only 15 words. Start with small paragraphs and then gradually increase the amount of text by adding larger paragraphs or several paragraphs for students to summarize with the same 15 word limit (Cunningham, 1982). Requiring students to summarize in writing using limited spaces to respond has been show to develop students' ability to identify levels of importance between main ideas and other information (Brown & Day, 1983).

K and ELL Scaffolding:

Kindergarten and ELL students may need the teacher to read a sentence or two to them out loud, and then the students can select three or four words that highlight the main subject or ideas. Teachers can work with the class or small group and write the summary on the board for the students.

Why Writing Summaries Works

The process of summarizing is a meta-cognitive process where students break information down into its essential parts and then evaluate and reflect on the information. When asked to write down their summaries, students synthesize the information and they are able to then better understand and own the information. Rosenshine and Stevens (1986) encourage summarizing that emphasizes an analysis of the information; developing this skill can have a powerful effect on comprehension and learning.

- Summarizing gives teachers and students the ability to monitor comprehension of information.
- Summarizing helps students understand the organizational structure of lessons or texts.
- Summarizing is an essential skill that is extremely practical and useful in future learning and work.

- Summarizing helps students to become better writers.
- Summarizing helps students analyze, reflect, synthesize, and evaluate.

Let students know that they summarize everyday as they share with a friend their favorite movie show or sporting event. Summarizing a full-color movie or an actual game that they participated in can be easy to share. Summarizing the verbal comments of new information or the written text of others can be a more challenging venture. "What I did last summer" is a basic summary of events.

Progress Monitoring for Writing Summaries

Through the process of writing summaries, our students will deepen their understanding of a passage of text. As they add their own thinking while consolidating and concisely organizing the information in their minds, they will become more conversant with the content. Written summaries help our students to infuse their own ideas together with the key concepts of a passage, and in the process, they develop greater comprehension and increased ownership of the information.

- Check to see if students begin to quickly identify main ideas from text.
- Identify if students can learn to write a summary in 15 words or less.
- Evaluate students' abilities to synthesize information concisely.

Summing It Up

Writing is an important process for developing expressive literacy. Students need interventions to ensure that they develop basic writing skills. If you can think it, then you can write it. Shared writing helps students because they have good writing modeled for them, and they can engage in the writing process together with their classmates. Make 'n Break writing can work for students at the most basic level of the writing process, since the words are already provided and they are just determining the order and organization of conveying ideas through writing. Summarizing is an important skill that will benefit students as they can synthesize information and then convey meaning concisely. Writing connections can help students see connections between the reading process and the writing process. As students develop their writing skills and engage in the writing process, they will become more confident communicators.

Reflection

1. How often do you provide your K-2 students writing opportunities?
2. How many strategies do you use to help students with their writing?

7

Integrating and Implementing Intervention Strategies

*"Learning is a treasure that will
Follow its owner everywhere."*

—Chinese Proverb

Tammy now comes to school with a big smile on her face and she looks forward to learning every day. It hasn't always been this way. When she first entered Willow Woods Elementary, she lacked confidence as a learner and avoided participating in lessons. Yet her first-grade teacher Mrs. Morgan spent extra time with her and several of her classmates. At times she felt a little self-conscious about being pulled to the back of the room to the small-group table, yet the activities were fun and very helpful. She appreciated the extra time, attention, and practice she received, since it helped make sense of things in her mind. It definitely meant a lot to her. Tammy could feel her understanding soar and it she knew that she was finally getting it. She had broken the code to words. Now a whole new world opened up to her. She was unsure how to thank Mrs. Morgan for all of her help, yet she would occasionally stay in at recess to see if her teacher needed any errands done. She liked being around Mrs. Morgan, because she finally felt confident as a learner. Tammy is now a confident learner who benefited from classroom intervention strategies.

It is important that we integrate and implement intervention strategies that work into our daily activities. As our friend and colleague Brad Wilcox says, "A little bit in every day, is better than a lot in May." If we will consistently add one more engaging strategy to our daily routine on a regular basis, we will soon expand our intervention repertoire. We will be prepared to meet the demands that our diverse students bring with them to school. The great thing about intervention strategies that work is they can be applied to a variety of content areas. The reading, writing, speaking, and listening strategies can be used in language arts, science, math, social studies, and a variety of other subjects.

Intervention

Our educational system needs an intervention if it is to truly educate a populace so that it can fully participate in the freedoms of being American. Without an education, in this global economy, unsuccessful students lose choices for the future; they literally become less free. If we are successful educating every child to their highest ability, they will enjoy access and the right to engage in the life of choice. Intervention needs to start as early as possible, especially for our students who come to school from a low socioeconomic background. Richard Weissbourd, an education researcher, articulates the need to realize that not all high-needs children are disruptive; many high-needs youth have "quiet" problems that often fly under their schools' radars. "The range of these problems is vast." Our young students in K-2 need and deserve intervention strategies that work. The preceding chapters provided you with 25 power-packed strategies that you can use in Tier I and Tier II intervention activities. These strategies help students because they increase the level of access, engagement, structure, and meaning that they definitely need to succeed.

Adding Arrows to Your Arsenal

As we add more intervention strategies to our instructional quiver, we will be prepared with the arrows in our arsenal that we need to help all students be effective learners. Every teacher needs to expand their instructional repertoire. A teacher's instructional repertoire is their quiver that holds the arrows that can get at the heart of student learning. The arrows are the teacher's intervention strategies that they use to increase student learning. Some arrows, or intervention strategies, will work with some students while other arrows, or intervention strategies, will work with other students. The key is to have enough intervention arrows in our arsenal to meet the needs of all our students. If we have a variety of intervention strategies that we can effectively direct toward learning targets, then we will be able to redirect the learning success for our students.

Access Many students face roadblocks to their education because they have challenges that limit their access to information. Language and literacy experiences are the key issues that affect a student's ability to access the important information provided at school. For many, the language issues arise because English is their second language. For other students, the language issues occur because students come from low socioeconomic backgrounds. Poverty's dramatic consequences affect language and learning.

Engagement Students direct their attention and their energy to things that are engaging. As they interact with information in ways that are engaging, they are more likely to extend and elaborate their learning. Most engagement requires an emotional response within the learning process. Hooking students to learning is the primary task of sustaining an engaged classroom.

Structure The key to instruction is developing the internal structures within students so that they can participate successfully. Some students naturally have the internal resources to create these structures or they have support at home to help create these structures. At the same time, many students lack the understanding of how to develop structures that will benefit them now and in the future. Every student needs the strategies that support an academic foundation and also requires the structures that will help them build an academic framework of learning.

Meaning Students develop meaning in two ways. They develop meaning externally by negotiating with others about what things mean. As students listen, read, speak, and write, they are able to negotiate with their teachers, peers, and book authors regarding meaning and their own personal understanding. At the same time, students develop meaning internally by negotiating with new information they receive. As students process new information and monitor and adjust their learning, the new meaning affects their worldviews. Learning changes people. A child in third grade learning the multiplication table becomes a different student when the math facts are known to a high level of automaticity. Learning helps us evolve in our understanding of how the world operates and how we might fit into it successfully.

Without intervention strategies that work, many of our students will accumulate failure and will become high-needs students. The number of these students dropping out of school is around 1.2 million each year. When we effectively use intervention strategies that work for our K-2 students, then we can literally change the trajectory of their results. Learning outcomes and life outcomes can be altered dramatically in positive ways. We can literally change the trajectory of a child's life by ensuring their success at school.

Repeated Practice with Intervention Strategies

Our students need to engage in intervention strategies at least 6 to 12 times for them to become proficient in the strategy. Glass, McGaw, and Smith (1981) emphasize that new strategies benefit students most when the students go through all five stages of learning.

1. Acquisition
2. Proficiency and Fluency
3. Maintenance
4. Generalization/Transfer
5. Adaptation

When students get to the fifth level of adaption, then they own the strategy and they can use their newfound knowledge and skills effectively in a variety of learning situations.

Professional Learning Communities and RTI

Professional learning communities (PLCs) are an important means to structure and support educational reform, organizational learning, and the use of data effectively. To do this, teachers need to be in a collaborative process or a professional learning community (DuFour, 2003). Working and learning with colleagues helps teachers develop the skills and abilities to use and apply data. Each member of the collaborative team contributes to dialogues, shares real-time experiences, and plans for the high-needs students. This "real-world" lens helps teachers interpret data, validate findings, and make planning decisions. The PLC allows teachers to collaborate and develop interim and formative assessments. Assessment literacy, alignment to standards, decision-making processes, data skills, and strategy identification are the strategies used on a collaborative team.

School-level mechanisms can provide incentives to support the PLC process including support for data inquiry, instructional improvement, and continuous improvement. Recognition of hard work can prompt the team to learn and grow. DuFour and Eaker (1998) recommend increasing capacity, supporting implementation, and creating sustainability to foster student achievement through the use of professional learning communities.

Increasing Capacity

Capacity is reflected as the organization's ability to learn by coordinating human, social, organizational, and structural capital and by making effective use of data. Features of human capital include knowledge, commitment, and disposition of school reformers. Social capital relies on professional networks, trust, and collaboration. Organizational and structural capital refers to the local organization's capacity to mobilize resource such as time, staffing, and materials for implementing change.

Supporting Implementation

Given the opportunity, teachers can develop the capacity to implement new strategies and reframe instruction, if they understand what needs to be done. The following steps can be taken to ensure implementation effectiveness:

- Describe the strategy in plain terms that everyone can understand.
- Describe the series of steps to take in order to use the strategy successfully.
- Explain how the strategy addresses the objective.
- Engage students in the strategy 6 to 12 times so that they can become proficient in the strategy.
- Describe results in understated terms.

To be implemented well, it can benefit students if they prioritize or rank instructional strategies to be implemented. After the ranking, determine which strategies will yield the greatest achievement gains and then choose two strategies to implement. Monitor and check for effect. Make sure that the new practices garners gains in achievement. If not, let go and try another strategy. NO ONE knows what will work for every child, so a certain amount of experimentation is necessary. However, use strategies that are science based and have high effect sizes.

Creating Sustainability

After the literature review of strategies to meet the needs of our students and the analysis of area achievement data, the characteristics discerned to be consistent with student achievement in high performing schools were organized into nine discrete categories of district and site leadership with an emphasis on the following:

1. High expectations
2. Clear goals
3. High levels of collaboration coupled with communication
4. Aligned curriculum to standards and assessments
5. Focused professional development
6. Teacher efficacy to include interventions
7. Adjustments of instruction
8. Site leadership as the facilitator to the overall work
9. Support of teachers by removing the barriers that impede teaching

All of these are interconnected to each other by accountability, which is the nucleus.

Summing It Up

Our classrooms are filling up with more and more students who struggle to keep up with core instructional lessons. These students need intervention strategies that will help them develop the skills and become proficient at the processes that lead to grade level learning. English Language Learners (ELLs), students from poverty (Title 1), and struggling readers need interventions that will help them grasp the skills and engage in the strategies that will help them succeed academically. As teachers, we need to expand our instructional repertoires. Students need interventions that will provide them access, engagement, structure, and meaning. As we add strategic arrows to our intervention quivers, then we will be better prepared to meet all of the demands that students bring to our classrooms. Students need help developing their listening, reading, math, speaking, and writing strategies. We hope that you have enjoyed all of the strategies contained in the book. As you regularly implement these interventions in your classroom, we know that you will enjoy the strategies and your young K-2 students will appreciate the support.

Reflection

1. Which intervention strategies will you implement in your classroom to make sure every student succeeds?
2. How will you share these interventions with your fellow teachers and colleagues?

References

Alliance for Excellent Education. (2009). Understanding High School Graduation Rates in the United States. Retrieved September 3, 2010, from http://www.all4ed.org/files/National_wc.pdf

Allington, R. (2010). Responding to RTI. *Education Week Teacher PD Sourcebook, 3(2).*

Archer, A. (2003). Vocabulary Development. Retrieved September 3, 2010, from http://www.fcoe.net/ela/pdf/Vocabulary/Anita%20Archer031.pdf

Bearne, E., Dombey, H., & Grainger, T. (2003). *Classroom interactions in literacy.* New York, NY: McGraw-Hill International.

Bell, J., & Bell, M. (2007). *Everyday mathematics.* New York, NY: Wright/McGraw Hill.

Blevins, W. (2002). *Building fluency: Lessons and strategies for reading success.* New York, NY: Scholastic Inc.

Bray, B. (2007). *Phonemic awareness activities and games for learners: Early childhood.* Huntington Beach, CA: Shell Education.

Brophy, J., & Good, T. (1986). Teacher behavior and student achievement. In M. Wittrock (Ed.), *Handbook of research on teaching, 3rd ed.* (pp. 328–375). New York, NY: Macmillan.

Brisk, M., & Harrington, M. (2000). *Literacy and bilingualism: A handbook for all teachers.* Mahwah, NJ: L. Erlbaum & Associates.

Brown, A., & Day, J. (1983). Macrorules for summarizing texts: The development of expertise. *Journal of Verbal Learning and Verbal Behavior, 22,* 1–14.

Brown-Chidsey, R., Bronaugh, L., & McGraw, K. (2009). *RTI in the classroom: Guidelines and recipes for success.* New York, NY: Guilford Press.

Burns, M. (2004). Empirical analysis of drill ratio research: Refining the instructional level of drill tasks. *Remedial & Special Education, 25(3),* 167–173.

Burns, M., Griffin, P., & Snow, C. (1999). *Starting out right: A guide to promoting children's reading success.* Washington, DC: National Academic Press.

California Department of Education. (2007). *Reading/language arts framework for california public schools: Kindergarten through grade twelve.* Sacramento, CA: CDE Press.

Cain, K., & Oakhill, J. (2007). *Children's comprehension problems in oral and written languages.* New York, NY: Guilford Press.

Caldwell, J., & Leslie, L. (2005). *Intervention strategies to follow informal reading inventory assessment: So what do I do now?* Columbus, OH: Allyn & Bacon.

Calhoun, E. (1999). *Teaching beginning reading and writing with the picture inductive model.* Alexandria, VA: ASCD.

Chapin, S., & Johnson, A. (1997). *The partners in change handbook: A professional development curriculum in mathematics.* Boston, MA: Boston University.

Charlesworth, R. (2004). *Experiences in math for young children.* Florence, KY: Cengage Learning.

Cooper, J., Kiger, N., & Au, K. (2008). *Literacy: Helping students construct meaning.* Florence, KY: Cengage Learning.

Cunningham, J. (1982). Generating interactions between schemata and text. In J. Niles and L. Harris (Eds.), *New inquiries in reading research and instruction, thirty-first yearbook of the national reading conference.* (pp. 42–47) Washington, D.C.: National Reading Conference.

Dolezal, S., Welsh, L., Pressley, M., & Vincent, M. (2003). How nine third-grade teachers motivate student academic engagement. *The Elementary School Journal, 103,* 239–267.

Evans, M. & Sorg, L. (2007). *Daily academic vocabulary, grade 4.* Monterey, CA: Evan-Moor.

Duffy, G. (2009). *Explaining reading: A resource for teaching concepts, skills, and strategies.* New York, NY: Guilford Press.

DuFour, R. (2003, May). Building a professional learning community. *The School Administrator,* 13–18.

DuFour, R., & Eaker, R. (1998). *Professional learning communities at work: Best practices for enhancing student achievement.* Alexandria, VA: ASCD.

Fisher, D., & Frey, N. (2007). *Better learning through structured teaching: A framework for the gradual release of responsibility.* Alexandria, VA: ASCD.

Fullan, M. (2001). *The new meaning of education change, 3rd edition.* New York, NY: Teachers College Press.

Geary, D., Bow-Thomas, C., & Yao, Y. (1992). Counting knowledge and skill in cognitive addition: A comparison of normal and mathematically disabled children. *Journal of Experimental Child Psychology, 54,* 372–391.

Gelman, R., & Meck, E. (1983). Preschoolers counting: Principles before skill. *Cognition, 13,* 343–359.

Geluykens, R., & Kraft, B. (2008). *Institutional discourse in cross-cultural contexts.* Muenchen, Germany: Lincom Europa.

Gentry, R. (2006). *Breaking the code: The new science of reading and writing.* Portsmouth, NH: Heinemann.

Gersten, R. & Baker, S. (2001). Teaching expressive writing to students with learning disabilities: A meta analysis. *The Elementary School Journal, 101,* 251–272.

Gersten, R., & Chard, D. (1999). Number sense: Rethinking arithmetic instruction for students with mathematical disabilities. *Journal of Special Education, 33(1),* 18- 28.

Girard, V. (2005). *Project Ella.* San Francisco, CA: WestEd.

Glass, G.V., McGaw, B., & Smith, M.L. (1981). *Meta-analysis in social research.* Beverly Hills, CA: Sage.

Glover, T., & Vaughn, S. (2010). *The promise of response to intervention: Evaluating current science and practice.* New York, NY: Guilford Press.

Grimes, S. (2006). *Reading is our business: How libraries can foster reading.* Chicago, IL: ALA Editions.

Hart, B., & Risley, T. (2003). The early catastrophe: The 30 million word gap by age 3. *American Educator, 22,* 4–9.

Hawley, W., Rosenholtz, S., with Goodstein, H., & Hasselbring, T. (1984). Good schools: What research says about improving student achievement. *Peabody Journal of Education, 61(4),* 1–178.

Heibert, E. (2009). *Reading more, reading better.* New York, NY: Guilford Press.

Henn-Reinke, K., & Chesner, G. (2006). *Developing voice through the language arts.* Thousand Oaks, CA: Sage Publishing.

Hirsch, E. (2003, Spring). Reading comprehension requires knowledge of words and the world: Scientific insights into the fourth-grade reading slump and the nation's stagnant comprehension scores. *American Educator, 27(1),* 10–29.

Hirsch, E. (2006). The case for bringing content into the language arts block and for a knowledge-rich curriculum core for all children. *American Educator, 30(1),* 8–17.

Honig, B., Diamond, L., & Gutlohn, L. (2001). *Teaching reading sourcebook for kindergarten through eighth grade.* Novato, CA: Arena Press.

Hoy, W., & Miskel, C. (2002). *Theory and research in educational administration.* Charlotte, NC: Information Age Publishing.

Irvin, J., Meltzer, J., & Dukes, M. (2007). *Taking action on adolescent literacy: An implementation guide for school leaders.* Alexandria, VA: ASCD.

Ivey, G., & Fisher, D. (2006). *Creating literacy-rich schools for adolescents.* Alexandria, VA: ASCD.

Johnson, E. (2009). *Academic language academic literacy: A guide for K-12 educators.* Thousand Oaks, CA: Corwin Press.

Lesh, R., Post, T., & Behr, M. (1987). Representations and translations among representations in mathematics learning and problem solving. In C. Janvier,

(Ed.), *Problems of representations in the teaching and learning of mathematics* (pp. 33–40). Hillsdale, NJ: Lawrence Erlbaum.

Lyon, R. (2001, March 8). *Using assessments and accountability to raise student achievement before the house committee on education and the workforce, subcommittee on education reform: 107th Cong. Testimony of G. Reid Lyon*. Retrieved from http://edworkforce.house.gov/hearings/107th/edr/account3801/lyon.htm

Maanum, J. (2009). *The general educators guide to special education.* Thousand Oaks, CA: Corwin Press.

Marzano, R., Pickering, D., & Pollack, J. (2001). *Classroom instruction that works.* Alexandria, VA: ASCD.

Mellard, D., & Johnson, E. (2007). *Practitioner's guide to implementing response to intervention.* Thousand Oaks, CA: Corwin Press.

Minton, L. (2007). *What if your ABC's were your 123's?: Building connections between literacy and numeracy.* Thousand Oaks, CA: Corwin Press.

Moats, L. (2000). *Speech to print: Language essentials for teachers.* Baltimore, MD: Brookes.

Muschla, J., Muschla, G., & Muschla, E. (2010). *Math teacher's survival guide: Practical strategies, management, techniques, and reproducibles for new & experienced teachers, grades 5–12.* Hoboken, NJ: John Wiley & Sons.

Nagy, W., & Anderson, R. (1984). How many words are there in printed English? *Reading Research Quarterly, 19*, 304–330.

Orfield, G., Losen, D., & Wald, J. (2004). *Losing our future: How minority youth are being left behind by the graduation rate crisis.* Cambridge, MA: The Civil Rights Project at Harvard University.

Pinnell, G., & Scharer, P. (2003). *Teaching for comprehension in reading: Grade K-2.* New York: Scholastic Inc.

Posamentier, S., & Krulik, A. (2009). *Problem solving in mathematics: Grade 3–6 powerful strategies to deepen understanding.* Thousand Oaks, CA: Corwin Press.

Pressley, M., & Billman, A. (2007). *Shaping literacy achievement: Research we have, research we need.* New York, NY: Guilford Press.

Rathvon, N. (2008). *Effective school interventions: Evidence-based strategies for improving student outcomes, second edition.* New York, NY: Guilford Press.

Restivo, S., van Bendegem, J., & Fischer, R. (1993). *Math worlds: Philosophical and social studies of mathematics and mathematics education.* Albany, NY: SUNY Press.

Riccomini, P., & Witzel, B. (2009). *Response to intervention in math.* Thousand Oaks, CA: Corwin Press.

Rose, J. (2006). *Independent Review of the Teaching of Early Reading.* Retrieved from http://www.standards.dcsf.gov.uk/phonics/report.pdf

Rosenfield, S., & Berninger, V. (2009). *Implementing evidence-based academic interventions in school.* New York, NY: Oxford University Press US.

Rosenshine, B. (1997). In J.W. Lloyd, E.J. Kameanui, and D. Chard (Eds.), *Issues in educating students with disabilities.* (pp. 197–221). Mahwah, N.J.: Lawrence Erlbaum.

Rosenshine, B., & Stevens, R. (1986). Teaching functions. In M. Wittrock (Ed.), *Handbook of research on teaching, 3rd ed.* New York, NY: Macmillan.

Routman, R. (1994). *Invitations: Changing as teachers and learnerss K-12.* Portsmouth, NH: Heinemann.

Rubin, J. (1994). A review of second language listening comprehension research. *The Modern Language Journal, 78(2),* 199–221.

Shapiro, M. (2009). *Mega fun math games and puzzles for the elementary grades: Over 125 activities that teach math facts, concepts, and thinking skills.* Hoboken, NJ: John Wiley & Sons.

Shiro, M. (1997). *Mega-fun math games and puzzles for the elementary grades.* San Francisco, CA: Jossey-Bass.

Shores, C., & Chester, K. (2008). *Using RTI for school improvement: Raising every student's achievement.* Thousand Oaks, CA: Corwin Press.

Soto-Hinman, I., & Hetzel, J. (2009). *The literacy gaps: building bridges for english language learners and standard english learners.* Thousand Oaks, CA: Corwin Press.

Snow, C., Burns, M., & Griffin, P. (Eds.). (1998). *Preventing reading difficulties in young children.* Washington, DC: National Academy Press.

Stone, C., Silliman, E., Ehren, B., Appel, K. (2005). *Handbook of language and literacy: Development and disorders.* New York, NY: Guilford Press.

Strickland, D., Ganske, K., & Monroe, J. (2002). *Supporting struggling readers and writers: Strategies for classroom intervention 3–6.* Portland, ME: Stenhouse Publishers.

Strong, J., Tucker, P., & Hindman, J. (2004). *Handbook for qualities of effective teachers.* Alexandria, VA: ASCD.

Sum, A., Khatiwada, N., McLaughlin, J., & Palma, S. (2008). *The Collapse of the national teen job market and the case for an immediate summer and year round youth jobs creation program, prepared for the United States house of representatives, subcommittee on labor, health, human services and education, washington, DC.* Boston, MA: Center for Labor Market Studies at Northeastern University.

Taylor, R. (2006). *Improving reading, writing, and content learning for students in grades 4–12.* Thousand Oaks, CA: Corwin Press.

The Learning First Alliance. (1998). *Every child reading: An action plan of the learning first alliance.*

Thomas, H. (2009). *English language learners and math.* Charlotte, NC: Information Age Publishing.
Troia,G., Shankland, R., & Heintz, A. (2010). *Putting writing research into practice: Applications for teacher professional development.* New York, NY: Guilford Press.
Van de Walle, J., & Lovin, L. (2006). *Teaching student centered mathematics: Grades K-3.* Boston, MA: Allyn & Bacon.
Vygotsky, L. (1978). *Mind in Society: The development of higher psychological processes.* Cambridge MA: Harvard University Press.
Walker, B., & Rivers, D. (2003). *Supporting struggling readers.* Ontario, CA: Pippin Publishing Corporation.
Wall, E., & Posamentier, S. (2006). *What successful math teachers do pre K-5: Research based strategies for the standards-based classroom.* Thousand Oaks, CA: Corwin Press.
Wilhelm, J. (2002). *Action strategies for deepening comprehension.* New York, NY: Scholastic.
Wilms, J. (2003). *Student engagement at school – sense of belonging and participation: Results from PISA 2000.* Paris, France: Organisation for Economic Co-operation.
Wood, T., & Turner-Vorbeck, T. (2001). *Extending the conception of mathematics teaching.* In T. Wood, B.S.Nelson, & J. Warfield (Eds.), *Beyond classical pedagogy: Teaching elementary school mathematics.* Mahwah, NJ: Lawrence Erlbaum.
Wixson, K., Lipson, M., Scanlon, P., & Anderson, K. (2010). *Successful approaches to RTI: Collaborative practices for improving K-12 literacy.* Newark, DE: International Reading Association.
Wolfram, W., Adger, C., & Temple, D. (1999). *Dialects in school communities.* London, UK: Psychology Press.
Wormeli, R. (2004). *Summarization in any subject: 50 techniques to improve student learning.* Alexandria, VA: ASCD.
Wyle, R., & Durrell, D. (1970). Teaching vowels through phonograms. *Elementary English, 47,* 787–791.
Zamel, V., & Spack, R. (1998). *Negotiating academic literacies: Teaching and learning across languages and cultures.* London, UK: Psychology Press.